Pilots' Weather

Pilots' Weather

A FLYING MANUAL

Ann Welch

A World of Books That Fill a Need

Frederick Fell Publishers, Inc.,
New York, N.Y. 10016

All drawings and photographs are by the Author except for the NASA satellite pictures.

Library of Congress Cataloging in Publication Data

Welch, Ann Courtenay Edmonds, 1917-
    Pilots' weather.

    Includes index.
    1. Meteorology in aeronautics. 2. Weather
forecasting. 3. Clouds. I. Title.
TL556.W39 1977   629.132'4   77-4991
ISBN 0-8119-0288-9

For information address:

Frederick Fell Publishers, Inc.
386 Park Avenue South
New York, New York 10016

First Published in England by John Murray (Publishers) Ltd.

PRINTED IN THE UNITED STATES OF AMERICA

0-8119-0288-9

To John Neilan:
test pilot and gliding pioneer,
through forty years of
British weather

Contents

Foreword

I am not a professional meteorologist, having learnt about weather from the ground up – the hard way. I have, however, worked a great deal with meteorologists both at the receiving end as a pilot, and as a task setter in World and National Gliding Championships, and I am very grateful for the help that they have always given me, in particular Professor C. E. Wallington now Director of the Institute of Marine Studies at the University of New South Wales.

For help with the preparation of this book I am sincerely indebted to Tom Bradbury of the Meteorological Office, Charles V. Lindsay of the US Weather Bureau, and John Ward, Secretary of the General Aviation Safety Committee and Editor of its Flight Safety Bulletin. The satellite photographs are from the National Environmental Satellite Service of the US Oceanic and Atmospheric Administration.

A great deal of generous assistance has come from the Editor of and contributors to *Pilot*, the official journal of the AOPA Foundation which does so much for the safety and the encouragement of the private pilot all over the world; also from the Editors of *Weather*, *Sailplane and Gliding, Soaring, Swiss Aero Revue*; and from John Simpson, expert on sea breezes.

The passages with rules round them throughout the book are specific examples of what the weather can do and why the pilot must never take it for granted. Those in the pilots' own words are acknowledged at the foot of each extract. I am also grateful to the following contributors to the magazines mentioned above on whose work I have based other passages: *Weather* – W. A. L. Marshall, R. F. Williams, C. M. Stevenson, N. Rutter, J. A. Taylor, D. Storr, Carl S. Benson, J. Gentilli; *Aero Revue* – Jiri Forchgott; *Sailplane and Gliding* – Adam Zientik; *Volo a Vela* – Plinio Rovesti.

A.W.

Hurricane Debbie over the Caribbean. Venezuela is marked by
dotted lines at the foot of the photograph, with Santo Domingo
at left centre. Taken from Satellite Essa, 19 August 1969 at
1806 hrs.

Introduction—At home in the air

The air is not our natural habitat as it is that of the albatross or the eagle, so we cannot expect to have the instinctive knowledge possessed by birds. We know our way around the land all right, but here we mostly get by without any real concern for the weather. We more often than not take a coat or, when it is needed, an umbrella, and remember times of the radio and TV weather forecasts; but since our lives are spent largely indoors the majestic life of the sky frequently passes unnoticed. When we go on holiday, particularly if we sail, fly, or ski, and certainly if we camp, the weather becomes much more important and we realise how little we really know about it – but even if we judge it wrong this rarely means more than getting a bit wet or cold.

When starting to fly, however, we expect the freedom of an unfamiliar and three-dimensional element. All our lives we have been *in* the air on the ground, so there seems no reason to believe that being *up* in it could be all that different. The instructor quite quickly teaches us basic pilotage skills in weather that he knows to be suitable, but thereafter it is possible to continue to fly without gaining any real understanding of our new element. Several factors contribute to this situation. For a start theoretical meteorology seems far removed from our simple needs. When flying in increasingly dirty weather and wondering what to do about it, we are not in the least concerned about the rotation of the earth or the coriolis force, but simply whether there is going to be a big enough gap between cloud and ground just for us. Since there can be no rule book which will give the answer to this or any other weather problem met with in the air, we have to work out what to do for ourselves; even if told on the radio of a clear weather alternate airfield, it is still we alone that have to get there.

Another reason is that it is easy to lose interest after we have made a vain effort to get to grips with the subject. The weather that one is supposed to know about as a pilot always seems to be made more complicated than necessary. This is not only because some erudite questions about its structure and behaviour are still not fully answered, but because it is too often surrounded by a mumbo jumbo of geostrophic forces, tephigrams, and spidery whorls on synoptic charts, instead of being treated as the everyday matter that it really is. So we tend to learn about the weather in unrelated bits and the subject as a whole becomes difficult to understand. At school it was all Horse Latitudes and Tropics of Capricorn; the private pilot does fronts, the glider pilot concentrates on thermals, and the yachtsman lives with Viking, Fisher, Dogger and Sole. So we give up. After all, there is probably a nice forecaster not too far away, and if he gets it wrong, it's not our fault.

This is a pity. In real life everyone's weather problems are, in fact, much the same whether we farm, fish or fly; we are concerned that we do not get caught out by weather that is unsuited to what we want to do. This is particularly so with flying whether it be in aeroplanes, gliders or even balloons. It certainly makes us free of this new element, and it brings rewards that are inconceivable to those who have never taken themselves into the air; but being right inside the weather is not the same as seeing and feeling it pass by. The air is too powerful; its movements and behaviour are on a scale that totally dwarfs our activities, and in the face of some of them, like hurricanes, we are powerless. So both for our enjoyment and continued existence we need to find out more about the air, and not just hope to pick up snippets of knowledge as we go along. We do not have to regard weather as a textbook subject because it is present all the time. From the clouds that form and grow in their infinite variety, and change and decay, we can learn most of what we need to know. Their language can be learnt by keeping our eyes open, and noting, and comparing what we see with what others have seen and experienced, and with what the forecasters say. As weather behaviour becomes familiar we find more that is interesting or un-usual, and if we don't see it for a day, it is almost like missing an instalment. When we go to a different part of the world we meet new people – and new weather. Quite soon we find that the sky can be read, and that we are getting better value from forecasts, or that we can make our own. We develop the instincts and cunning that enable us to know how to thread our way safely through or between the excesses of the weather jungle. This book has one aim; to enable any pilot, or would-be pilot, to reach this stage as quickly and pain-lessly as possible.

ANN WELCH

Part 1　Weather analysed

1 Why we get weather

The basic causes of weather are simple, and we can base our future learning about the whole subject on just two fundamental rules.

(1) *When air at a certain pressure is warmed its density decreases, it becomes buoyant in relation to the surrounding air, and rises. Conversely the density of cooled air increases, and it sinks.*

(2) *When air is cooled enough some of the invisible water vapour that it carries condenses and creates cloud. When cloud is warmed it evaporates back into invisible water vapour.*

In practice we see the result of this second rule in a car. The condensation on the inside of the windscreen is just a thin film of cloud; when we put on the heater the condensation evaporates.

These two rules apply equally over those millions of square miles around the hot Equator, in every spell of rotten weather, and in the weakest little upcurrent that a glider pilot leaves in disgust. Obviously there will be different reasons for the air to go up or down, clouds will come in all sizes, and the terrain over which air moves will contribute its share of modifications, but these are details. They are fascinating and necessary to know but secondary to the main theme, *that whenever air is cooled sufficiently cloud will form*; and our weather is just clouds arranged in different ways. So we are concerned in this book with why clouds develop, what form they will take, what they will look like, and when they will go away.

The sky is thin

Because of the huge difference in warmth received from the sun at the Equator and at the Poles, and because the earth goes on spinning, it is obvious that the air in which all these clouds will form is being constantly churned around. This turmoil, with its ever changing temperatures, densities and pressures, and the humidity which the air possesses, is confined within a remarkably shallow layer. From the surface upwards the air thins out imperceptibly as height increases, but the depth with which we are concerned is about 50,000 ft, or a mere 10 miles thick. This is horizontally equivalent

Luck is useful too . . .

. . . Altitude 11,000 ft. The sun's rays on top of the perfectly solid creamy cloud beneath me made it very difficult to judge the distance between the plane and the cloud deck. All of a sudden, I found the clouds billowing up around the plane, rising more rapidly than the plane would climb. Within a matter of seconds, I was entirely involved in a sort of white wool, which 5 minutes later was black – but black, like a darkroom. I could barely make out the instrument panel.

There was turbulence within the cloud, so violent that, at times, I noted the plane climbing 2000 fpm, and a moment later plunging downwards at the rate of a thousand or more. There was no sense of control whatever in the plane, the same as if ailerons, rudder, and elevator had been disconnected. The compass spun one way and then the other. It was like being a peanut in a mixmaster. . . . all of a sudden out of the mist off my right wing, a fir tree whipped by. Five minutes later a cliff was momentarily visible close by my left wing. I had lost a great deal of altitude, had climbed up little by little, only to lower again, but by that time I knew with a certainty what was coming. A few seconds later, with a little cone-shaped tip of a mountain at my very nose, I was prepared. I cut the switch and the gas, and ploughed into the mountaintop. The altimeter marked 11,800 ft.

The plane ploughed forward into the ground, stopping within 7 or 10 ft. Pine saplings, which cut slots in the wings, broke the roll forward . . . the cabin was flooded with gasoline, and a mass of gasoline vapour rose over the hot motor. . . . I hopped out in one curve, like a fish out of a tank. . . . In a matter of moments I was sopping wet, the cold was something terrible, and I climbed back into the cabin still dripping with gasoline, to think it over.

William Spratling, on a flight from Iguala to Mexico City which he had previously made 2000 times.

—AOPA Pilot

to just 7 minutes driving on a motorway; a jet fighter could go straight upwards through it within 5 minutes of take-off. In light aeroplanes and gliders we mostly use little more than the bottom tenth of this air – which holds the worst of the weather. Passenger jets usually have two-thirds of the atmosphere below them and nine-tenths of the cloud, so the lucky people are above most of it most of the time. It is, of course, the meagre depth of our atmosphere that determines the scale of weather – the size of depressions or the height of storms. If the atmosphere were deeper there would probably be larger areas of bad weather further apart, and if it were

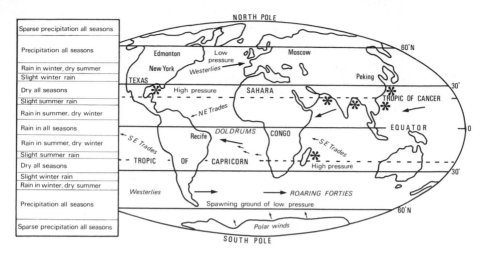

Sparse precipitation all seasons	
Precipitation all seasons	
Rain in winter, dry summer	
Slight winter rain	
Dry all seasons	
Slight summer rain	
Rain in summer, dry winter	
Rain in all seasons	
Rain in summer, dry winter	
Slight summer rain	
Dry all seasons	
Slight winter rain	
Rain in winter, dry summer	
Precipitation all seasons	
Sparse precipitation all seasons	

1.1 World weather

In the northern summer there is a seasonal shift of weather belts to
the north, and in the southern summer to the south. The asterisks
show hurricane, typhoon, and cyclone trouble spots.

thinner, they would be smaller but closer together. In a trans-
atlantic jet we may fly over a single depression for 600 miles, but
nowhere is it more than 6 miles thick, or deep. This is not the
impression that one retains from most diagrams of atmospheric
circulations, which too often give a ludicrously exaggerated vertical
scale; if the earth's diameter were the height of this page, the
atmosphere would be as thick as only 5 sheets.

This skinny covering of air is all that effectively shields the earth
from the many billion horse-power of heat energy that it intercepts
from the sun; heat that is absorbed by the ground, or reflected from
the clouds, and which is the starting point of all the interrelated
processes that determine our weather. We now need, therefore, to
look at its ingredients in a little more detail, even at the price of
some dull workaday pages. If we do not have clear in our minds a
few essential points about the structure of the atmosphere it will be
less easy to achieve any real understanding of how and why weather
changes take place, and the flying-orientated parts of this book will
be of less use.

Density and pressure

We must start by being clear about the difference between density and pressure. The *density* of the air is the closeness together of air particles, and is expressed in mass per unit volume. For example, at 15°C a cubic metre, at sea level, contains 1226 grammes of air.

The *pressure* is the weight of all the air above any given unit area of the earth's surface. The total weight of air above any square foot of surface is about 15 lb. As we go higher there is less air above, and so the pressure is less. On top of Mt Everest, almost 30,000 ft, the pressure is only about one-third that at sea level – about 5 lb/sq ft.

Although pressure is strictly a force per unit area, it is sometimes expressed in inches of mercury (the height of a column of mercury

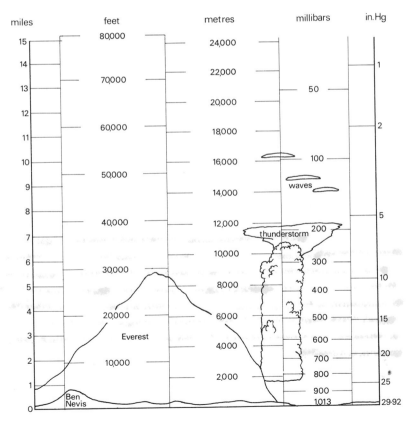

1.2

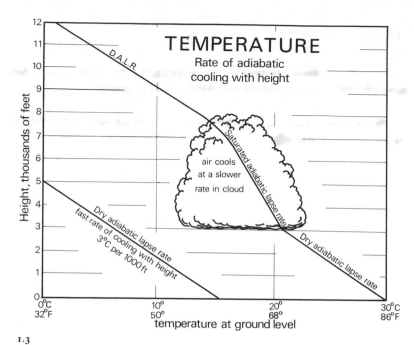

TEMPERATURE
Rate of adiabatic
cooling with height

D.A.L.R

Saturated adiabatic lapse rate

air cools
at a slower
rate in cloud

Dry adiabatic lapse rate
fast rate of cooling with height
3°C per 1000ft

Dry adiabatic lapse rate

Height, thousands of feet

0°C 32°F | 10° 50° | 20° 68° | 30°C 86°F

temperature at ground level

1.3

which produces the same pressure). In aviation it is more usual to use an international unit of pressure called a *millibar*, or *mb* for short. The distribution of pressure in the atmosphere is the basis on which weather forecasts are made and presented.

Pressure and temperature

We notice changes of temperature because we feel hot or cold, but unless we do something less usual, like blow up a bicycle tyre, we may forget that the temperature of the air alters whenever its pressure changes. As pressure is increased the temperature will increase – as with the hot cycle pump, and as pressure is reduced so the temperature will lower – a cause of carburetter icing. Alterations of temperature caused by changes in pressure are termed *adiabatic*. The word simply means that a parcel of air can be subjected to changes of pressure without it gaining or losing heat from, or to, outside sources.

To summarise this density, pressure and temperature inter-relationship; if our parcel of air at the surface is warmed, then its pressure will initially remain unaltered, but due to the higher temperature its density will be slightly less than that of the surrounding air. It becomes buoyant and so rises. As it rises its pressure will decrease – because there is now a lessening amount of air above it – and its temperature will likewise decrease because the change of temperature will be very nearly adiabatic. The amount by which it cools as its height increases is called the *adiabatic lapse rate*. If air is dry – that is, without cloud – the rate is 3°C per 1000 ft. If, however, the air is saturated as it is in cloud, the adiabatic lapse rate is lower, only 1·5°C per 1000 ft. This is because heat is given out by the processes of condensation, and it is known as the *saturated adiabatic lapse rate* (Fig. 1.3).

Many of the world's airfields are situated several thousand feet above sea level. The lower pressure at altitude adversely affects take off performance – due to the reduced engine power available combined with the need to reach a higher *true* airspeed to get off. If the air is very hot this will make matters worse as it is equivalent to increasing the effective altitude. (*See* Appendix 4, 'Density Altitude'.)

Heat

The heat of the sun, which is not reflected back into space by cloud or dust, or radiated away, reaches the earth and warms it. This warming will be uneven because different types of surface absorb heat at varying rates, and because of the periodic effect of night and day, and the seasonal inclination of the earth to the sun. In turn the warmed surface will warm the air in contact with it, also of course unevenly.

Wind

Wind is simply air moving from where the pressure is higher to where it is lower, just as air in a blown-up balloon wants to escape until the pressure inside equals that outside. Because temperature and pressure changes are related, it should not be forgotten that the cause of a wind may not be only a simple pressure change, but could also be due to air moving from a cooler to a warmer place. Within a large area of substantially uniform pressure, with consequently little incentive for air to move on a big scale, there will be winds created by local temperature differences, as between cool sea and warm land. There will, of course, be some small pressure difference between the two, but not enough to effectively modify the big pressure pattern. This is one reason why the wind often drops in the evening and at night – winds produced as a result of heating during the day cease, and the overall pressure change, or gradient, is insufficient to cause the air to move with any force. Lack of heating is also the reason that a fine winter day is often less windy than an equivalent fine summer day.

Winds do not blow *directly* from higher pressure to lower because they are deflected by the rotation of the earth. When blowing away from the fast-spinning equator they 'get ahead of themselves' as the speed of rotation slows towards the Poles. This curves the wind progressively – to the right in the Northern Hemisphere and to the left in the Southern.

Humidity

Air is never dry except in a laboratory. The atmosphere carries as invisible water vapour a fantastic quantity of moisture – a single thunderstorm alone can deposit some 10,000 tons of it in 5–10 minutes. As we saw in our two Rules, the amount of moisture that the air can carry depends on the temperature; *the warmer the air, the more moisture it can carry, and the cooler the air, the less.* So when air cools there comes a time when it is saturated and can no longer support all the water vapour that it holds. Any further cooling will result in some of the moisture condensing on to tiny nuclei in the air as droplets of water which remain suspended as cloud. They are minute, but in certain circumstances they will be able to grow bigger, and then will fall as rain.

The process of condensation liberates heat into the air; known as the *latent heat of condensation*. If sufficient moisture is present, condensation will continue taking place in the rising air, and cloud will go on forming.

When saturated air, or cloud, subsides or moves into warmer regions, so that the cloud droplets evaporate, work is needed to make this change. The energy required is taken out in the form of heat, and the temperature of the air is consequently lowered. The loss of heat due to evaporation is most apparent when we have to stand about in wet clothes, particularly if a wind is increasing the rate of evaporation. Humidity is the term for the dampness of the air. The amount of moisture that the air contains at any given moment is called the *absolute* humidity – or just the humidity, whereas the amount that it holds in relation to the maximum that

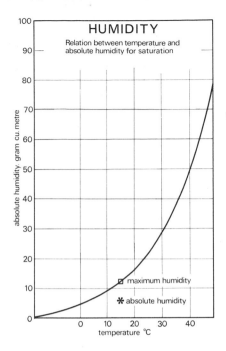

HUMIDITY
Relation between temperature and absolute humidity for saturation

absolute humidity, gram cu.metre

temperature °C

□ maximum humidity

✳ absolute humidity

1.4 *Example.* If at 15°C a cubic metre of air actually holds 6.5 grammes of water vapour the *absolute humidity* is 6.5 g/cu m (point ✳). As air at 15°C could carry 13 g/cu m (point □) the *relative humidity* of the sample is 50%.

it could contain for that temperature is called the *relative* humidity. For example, if a cubic metre of air at 15°C and 1013·2 mb pressure held 6·5 grammes of water vapour the absolute humidity would be 6·5 g/cu m. But at this pressure and temperature a cubic metre of air is able to hold 13 grammes of invisible water vapour, so if it actually holds only 6·5 grammes, the relative humidity is 50% (Fig. 1.4).

Dew point

If air is gradually cooled down, the time will come when it becomes saturated and some of the water vapour starts to condense; the temperature at which this occurs is known as the *dew point*.

A beautiful day . . .

The trip was to take 1 hour 10 minutes. I checked the weather Friday evening and was told a front moving in from the SW would be no problem. Saturday morning it was the same story. The weather people said it was all okay for the trip; so we were in the air by 0700, expecting

—continued

to be at our destination in slightly more than an hour. As we climbed out, heading south, I commented on what a beautiful day it was.

But 20 minutes south of Atlanta we ran into rain, light turbulence and low visibility. I called weather; the report was the same as before. They saw no reason for me to turn back and were optimistic. We continued, but within another 10 minutes conditions had deteriorated to the point that I elected to return. Atlanta, which an hour earlier had been covered only by a low, very thin, broken layer, was now socked in. I couldn't find a hole through which I could let down. I heard on the radio of 1000-ft ceilings, snow, and freezing rain – no place for a VFR pilot like myself. Atlanta radar, because of the weather, could not pick me up and advised my landing somewhere else.

The Mooney still had plenty of gas so we turned NE, thinking to out-run the edges of the front, and be on the ground in less than half an hour. But I was almost in serious trouble. A thousand feet below was the top of the cloud cover. Several thousand feet above was another cloud cover, and between the two, in the leading edge of the front, were scattered patches leading me to think they might meet soon, with me sandwiched between. I had no reference to the ground and could see nothing but the white glare of clouds. I kept a close eye on the altimeter and hoped the cloud deck below was not slanted.

In about 15 minutes we had left behind the white-out and were flying at 7500 ft in clear air atop an overcast stretching to the horizon. The Great Smokey Mountains could be seen over 100 miles away. I was on the radio almost constantly now, checking the weather and hearing the same story, ice, low overcast, snow. Where had it all come from? I proceeded on into South Carolina, but had now flown off my chart. Greer Radio told me the entire south-east was socked in. Then he came back to say I might be able to get into Bristol, up on the Virginia border. By now I had been in the air over 2 hours. I turned back to the NW on a course that would take me directly across the ominous 8000 ft-plus peaks of the Smokies. Greer Radio was now calling every few minutes. I knew the one thing that would mean the end of the ball game would be for me to lose my cool. I made a conscious effort to remain calm so my non-pilot friend would not be alarmed. But I need have had no worry. He was looking at the mountains, wondering aloud at their beauty and occasionally asking, 'What happens when we run out of gas?' or 'Is your insurance paid up?'.

As we crossed the mountains it was apparent that they had acted as a barrier in holding back the low ceilings accompanying the front. The valley beyond – where we were to land – was clear. But already clouds were beginning to wisp through the passes and wind down the north side of the ridges.

—Extract from account by Robert Coram. *AOPA Pilot*

The effect of land and sea

When trying to work out what the continually changing temperature, pressure, and humidity is doing to the weather, the source of the air and the sort of surface over which it has subsequently moved has to be taken into account. Oceans, land masses, and mountain ranges all modify the characteristics of air that flows over them – the Himalayas, after all, stick up through most of the atmosphere.

Air is affected by the type of surface over which it travels, the time that it spends over that surface, and the amount of heating or cooling that is imparted to it.

If air moves over the sea for any length of time it will pick up moisture; if the sea is cold it may cool the air sufficiently to cause cloud or fog. When air arrives over land after a long sea crossing it will be moist, so if it is forced to rise up over high ground it will expand and cool. The extent of this cooling will determine how much cloud will be produced over the hills. So if we are flying towards high land in bad weather we should expect more and lower cloud. If the hills are high enough the air will be substantially cooled as it passes over them, and rain as well as cloud will result. The rain will be left on the mountain, so that the air moving on beyond the high ground will be drier. When cool air moves over a large warm land mass it will become heated and its *relative* humidity will decrease, although unless rain falls its *absolute* humidity will stay the same (Fig. 1.5).

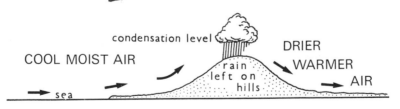

condensation level
COOL MOIST AIR
DRIER
WARMER
rain left on hills
AIR
sea

1.5

To anyone with an orderly mind the involved behaviour of the air must seem chaotic, with little prospect of ever discovering what state it will achieve next. Fortunately, the physical laws which govern the activities of air are exact, and exacting; so in any given set of circumstances, the air *will* behave in a certain way. This is why we are able to recognise and give names to weather patterns and systems, and forecast what they will bring.

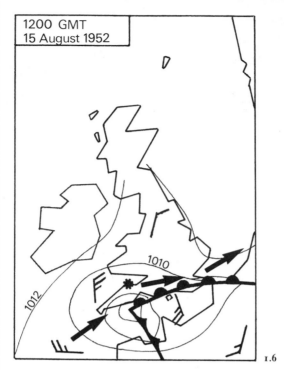

1200 GMT
15 August 1952

1.6

The Lynmouth disaster

Looking at this weather map the pilot in a hurry would be excused
if he thought that the bad weather would be just along the south
coast and up from Cherbourg. But the most disastrous rain and
floods occurred in North Devon, marked with a star. The
Lynmouth floods of 1952 were caused by a small depression to
the SW of England bringing unstable thundery air up from
France. This air drifted up across the Channel, across southern
England, swinging round to become a northerly wind blowing
towards the N coast of Devon. There had been a lot of rain in the
area over preceding days and on 15 August, but the critical fall
occurred during the late afternoon, continuing until midnight.
The situation at this time was not dissimilar to the Welsh floods
(page 71) in that the rain-filled and unstable wet air was
accelerated towards the coast during the afternoon as a result of
thunderstorm convection lifting air massively over the land. At
Lynmouth it hit was even higher, up to 1500 ft above sea
level within four miles, giving a powerful orographic effect. The
rain continued for 12 hours, 7 inches falling between 1700 hrs and
midnight. The disaster was not caused directly by this rain, but
resulted from its effect on the already boggy and saturated moors.
The underlying rock of Exmoor prevents water soaking away so
when the really heavy rain of August 15 arrived there was
nowhere it could go but down the face of the hills; something like
half a million tons of water per square mile did just this.

2 Outlining weather systems

The effect of the interplay of temperature, pressure, and humidity gives us all the weather we have, from the small and local patch of sea fog lying across our destination airfield for just a few hours to weather in the grand manner which may afflict us for weeks. In some parts of the world, such as the subtropics, where variations in temperature are neither particularly great nor sudden the weather is both stable and predictable. But where large masses of air of widely different temperature continually come in contact the weather that results will be both unstable and difficult to time. Needless to say, the British Isles and the Atlantic lie in such a zone, and the situation is further complicated by Britain being a rocky island at a 'cold' latitude lying in a relatively warm sea.

When such huge masses of air of different temperature come together they create the big pressure system well-known as the *Depression*, or *Low*. It has these names because the pressure within the system is lower than average, and it produces characteristic and sometimes formidable weather. The greater the temperature difference between the two air masses the more vigorous the depression will be, and the smaller the difference, the milder or weaker the depression that will develop. An agglomeration of air of higher pressure, and more uniform temperature, is known as an *Anticyclone* or *High*, and it gives more settled weather.

Before investigating the habits of Lows and Highs there is a semantic confusion which should be cleared up in order to make it easier to visualise what is actually taking place within these all-pervading and impressive air mass systems. To a pilot, thinking in three dimensions, the words Low and depression have a 'downward' sense, and High, even if not anticyclone, has an 'upward' meaning. As a result it may be difficult to visualise *a Low or depression as a region in which air is substantially ascending – that is why it has so much cloud; and a High as a region in which air is substantially subsiding*, which is why it develops relatively insignificant cloud or none at all. Lows and Highs were named at a time when low pressure and the storms to which it gave birth were of concern primarily to the seaman, and high pressure more often than not gave blessed relief. The barometer would fall as the Low approached, and rise as a High developed. But in flying we are concerned not only with the weather that is approaching, but with what the air is doing underneath us, and at high levels. Because it produces cloud we are just as concerned with the vertical movement of air as with the horizontal; when pressure is falling, it is doing so because air is expanding and going up. On its way up condensation takes place when the air cools sufficiently, and cloud, and perhaps rain, are produced. We can even draw a simplified diagram to get the picture quite clear in our minds (Fig. 2.1).

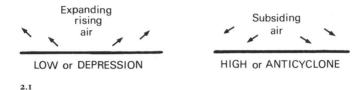

Expanding
rising
air

LOW or DEPRESSION

Subsiding
air

HIGH or ANTICYCLONE

2.1

The depression

A depression may develop anywhere, but the ones that we know and regard as typical in Europe, are most prolific between latitudes 50°–60° (in both hemispheres) (Fig. 1.1). This is the natural meeting place of the masses of cool air moving away from the polar regions, and those of warm air that have risen above the tropics and are moving towards the poles. On meeting, the warm air will stay above, or flow up over, the cold; in doing so it will expand, lower in pressure, be cooled, and produce vast quantities of cloud. The denser polar air stays underneath, moving in towards the area of lessening pressure created by the warm air upflow, and initiating the characteristic circular motion around the system. Because of the W–E rotation of the earth, in the Northern Hemisphere the direction of flow around a Low is anticlockwise; in the Southern Hemisphere it is clockwise (Fig. 2.2).

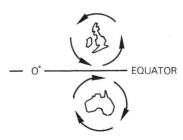

— 0° ———————— EQUATOR

2.2

A depression may be deep or shallow. It is deep if there is a steep pressure fall towards the centre – indicating that there is rapid expansion and rising of the air. This will produce strong winds, and a lot of cloud and rain. A shallow depression has a more gradual pressure fall, or gradient, towards the centre because the ascending movement is only gentle. There will be plenty of cloud, but less strong winds, and maybe not so much rain. When the pressure in the Low ceases to fall, the system will start to 'fill', weaken, and die away.

A very large depression may be as much as 2000 miles across, and a small one as little as 50 miles, but both can be equally fierce. Depressions may form one after another and follow a similar track, so that the weather remains unsettled for a week or more. Sometimes, in Britain the rain goes through on succeeding nights and the days are relatively clear, but too often the opposite seems to happen; it rains and blows all day, and the nights are starlit and calm. Typhoons and hurricanes are small extra-ferocious depressions with an unusually large pressure fall at the centre (page 232).

Fronts

Where the ex-tropical air and the ex-polar air come into contact there must obviously be a discontinuity of temperature and pressure over a large interface area. The actual temperature difference will depend on a lot of things, including how long the two lots of air have been on their journey, and whether it is summer or winter. Where the warm air is expanding, floating up, and producing cloud over the cooler air ahead of it, the discontinuity is known as the *warm front*. The *cold front* is where cold higher pressure air is advancing under warmer air, and creating cloud mainly by pushing it up. The fronts are drawn on weather maps as lines, but these show only where the change in temperature and pressure between the two lots of air lies at ground level. Above the ground frontal cloud and weather covers a much greater area, with maybe 100 miles or more of poor flying conditions ahead of the warm front. This region is mostly filled with superimposed layers of cloud, usually thick, and often running one into the other (Fig. 2.3).

Although the complete depression is moving along, the fronts do not necessarily move at the same speed as the whole system. The higher pressure cold front usually travels faster than the warm front

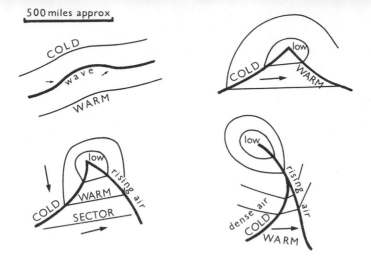

2.3

because it is being pulled towards the low pressure ahead. Gradually the cold front catches up with the warm, starting to do so where the two fronts are closest together at the centre. The overlapping fronts are said to be *occluding*. Eventually the *occlusion* may extend most of the way along the front line.

2.4 The occlusion weather will be of cold frontal or warm frontal type depending on the relative temperatures of the two lots of cold air.

The anticyclone

An anticyclone is a stable, slow moving, often vast, mass of air whose pressure is higher than 'average'. Unlike the Low, which is born in areas of the world where there is considerable variation in temperature, and humidity, such as the North Atlantic, the anticyclone will generally develop where there are more uniform conditions. The air in a High is relatively homogeneous, without fronts, and is generally, and gently, subsiding. Air sinks into it from above, and leaves as a divergent flow nearer the surface. Although the rate of subsidence is very gradual, it is enough to have a strong stabilising influence on the air through its compressive and warming effect.

The air circulates around an anticyclone in the opposite direction to a Low (Fig. 2.5). In the Northern Hemisphere the circulation is clockwise, and in the Southern Hemisphere, anticlockwise.

2.5

The areas of the world where anticyclones are most common are over the Poles and in the Horse Latitudes, or sub-tropics. The cold area anticyclones, including those over Siberia in winter, develop because cold dense air sinks and in the lower levels there are huge quantities of it doing so. The high-pressure belts of the sub-tropics exist as a result of the massive uplifting of tropical air over the equatorial regions. It is replenished by air on either side of these regions subsiding and moving in towards the equator. These belts of high pressure, which are around latitude 30° in both hemispheres, produce large and long-lasting anticyclones, including of course, the well-known Azores High.

Anticyclonic conditions will also develop more locally as a pressure balance mechanism, usually between two adjacent depressions. Such a small region of marked pressure rise is called a *ridge* (page 37). An anticyclone will start to decline when air ceases to sink into it at high levels. The central pressure drops and the whole circulation weakens. Decline also may occur if the system drifts over a quite different surface, such as from cold land to relatively warm sea.

The big picture

Because we constantly hear about depressions and anticyclones in forecasts, it is easy to get the idea that the weather is made up entirely of these two systems. The depression, in particular, features largely in our European lives because we are at the receiving end of one of the most active spawning grounds of the Low in the world – the North Atlantic. But since we may fly elsewhere and find quite a different situation, and since Europe does have other weather as well, we should always keep an open mind and consider whatever we find from first principles. That is, that *the atmosphere is a thin layer of damp and dirty air in a state of continuous movement, and that any change in temperature will alter the pressure, which in turn will alter the cloudiness.*

We need to remember that even more than our own cash accounts, the big pressure picture has to balance. There are times when pressure may be fairly uniform from Newfoundland to Kiev, but if deeper and deeper depressions start growing in the Atlantic, pressure will begin to get higher elsewhere. We should look out for it.

In the next chapters we will return to the big pressure systems to see how they appear on weather maps, and what they look like in the air. *But whatever weather situation is presented to us, our concern is primarily to consider how it* is *changing, and how it* will *change. If we ever leave a met office without having done our best to get answers to these questions, then we have not got a forecast.*

The Tynemouth balloon race

The flow around an anticyclone is well illustrated by the garden fête balloon race of the Tynemouth Priory Round Table on 16 May 1964. Nine-inch balloons filled with hydrogen were used with name and address labels hopefully attached. The race looked like being a failure when the balloons drifted in a monotonous stream straight out into the North Sea, borne on a light SW flow in a mainly cloudy airstream. But to everyone's surprise, that evening at 2045 hrs a balloon was found in Germany 400 miles away, having averaged a speed of 40 knots. By the next day there had been a succession of landings right down into Italy. It is even possible that some foundered in the Mediterranean.

The weather map (A) showing the surface plot for the day of the race is marked with some known landing places, but it does not give the whole answer. The clue is discovered in the upper air chart (B), which indicates a strong northerly flow, although it presupposes that the tiny balloons rose to 30,000 ft or more, into a temperature of $-40°C$ without bursting. Some similar balloons were quickly tested. They took about 2 hours to rise into the strong upper flow where the wind blew at 60–70 knots; they then burst, but took several hours to descend, all the time drifting on across Europe.

On this occasion the upper winds were strong, and over the Alps and Mediterranean quite different in direction to the low level winds. A flight across the Alps from S to N, for example, would have started and ended with light easterly winds, but a strong head wind would have been encountered at the height needed to cross the mountains with safe clearance.

—continued

The Tynemouth balloon race

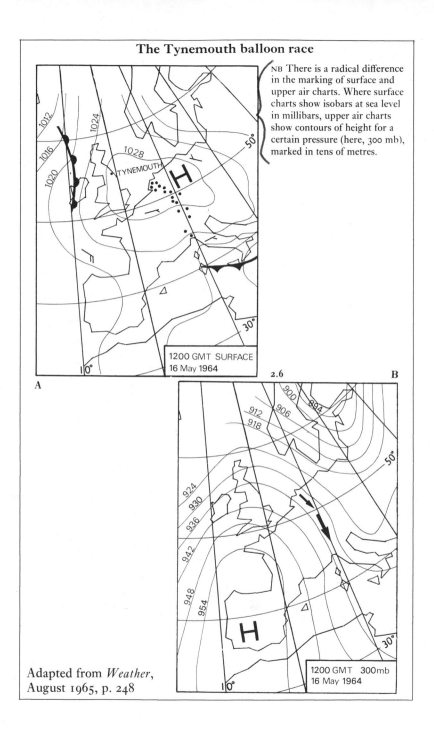

NB There is a radical difference in the marking of surface and upper air charts. Where surface charts show isobars at sea level in millibars, upper air charts show contours of height for a certain pressure (here, 300 mb), marked in tens of metres.

1200 GMT SURFACE
16 May 1964

2.6

A

B

1200 GMT 300mb
16 May 1964

Adapted from *Weather*,
August 1965, p. 248

3.1 Satellite picture of depression

An Atlantic depression sweeping over most of Britain, with Iceland
in the clear to the north. The area covered by the sheet of warm
frontal cloud is extensive and is likely to last over Scotland for
8 hours or more. Taken from Satellite Essa at 1348 GMT on
18 July 1969.

3 Signs, symbols, and systems

Because people everywhere want to know what is happening to the weather there is a need to be able to communicate meteorological information in an internationally comprehensible form. The most effective way is by means of a weather map, or *synoptic chart*, to name it correctly. This shows by symbols the behaviour of the air over the region concerned in terms of pressure distribution, and displays information about wind, temperature, cloud cover, and precipitation. So even if the met man speaks only Khatmandu we will still be able to understand his chart.

Weather maps appear daily in some newspapers and on television, some using the agreed international symbols for plotting the data, and others not. Some of these instead display a bewildering collection of signs from puff-cheek cherubs to umbrellas, in the hope that everyone will known that these mean wind and rain. They probably do, but for flying factual information is necessary, so that we can work out for ourselves the situation prevailing at the moment *and the changes that are likely to take place*. We do not want a third-hand digest which someone along the line has possibly biased to appease rained-out hoteliers. But be warned that even some of the better newspaper maps have inconsistencies, one having the wind speed in little circles, and another putting the temperature in them. As a result some pilots, either faced with cartoon charts or put off by a seemingly endless quantity of confusing symbols, never get around to finding out how to extract the maximum information from the weather map. They just rely on what the forecast or the forecaster says, because it seems a waste of time to attempt to do his job without either his experience or knowledge. But if we know what to look for, and have learnt the international symbols – which are not difficult – there is much more information obtainable from any weather map, even those quaintly presented, than appears at a quick glance. Self-help practice in digging out every scrap of information from charts will never be wasted. The aeroplane and the weather are a lot more mobile than the meteorologist or his office, and sooner or later the time will come when we will have to do the best with whatever information we can get – even from an old newspaper.

3.2 Weather symbols and their use

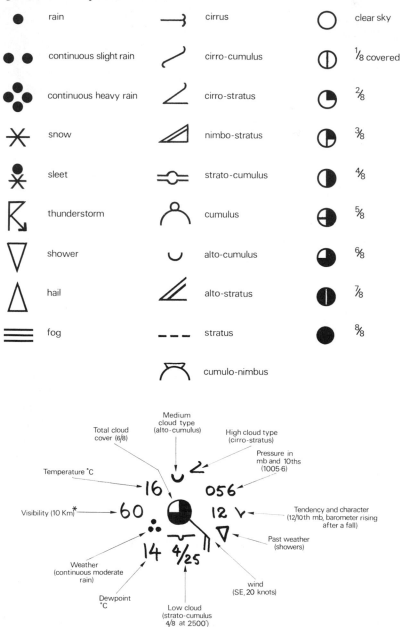

●	rain	⌐	cirrus	○	clear sky
● ●	continuous slight rain		cirro-cumulus		⅛ covered
	continuous heavy rain		cirro-stratus		²⁄₈
✳	snow		nimbo-stratus		³⁄₈
	sleet		strato-cumulus		⁴⁄₈
	thunderstorm		cumulus		⁵⁄₈
▽	shower	∪	alto-cumulus		⁶⁄₈
△	hail		alto-stratus		⁷⁄₈
☰	fog	- - -	stratus	●	⁸⁄₈
			cumulo-nimbus		

Total cloud cover (6/8)

Medium cloud type (alto-cumulus)

High cloud type (cirro-stratus)

Temperature °C

Pressure in mb and 10ths (1005·6)

Visibility (10 Km)*

Tendency and character (12/10th mb, barometer rising after a fall)

Past weather (showers)

Weather (continuous moderate rain)

wind (SE, 20 knots)

Dewpoint °C

Low cloud (strato-cumulus 4/8 at 2500')

16   ∪ ⌐   056

60 · ⬕ 12 ▽

·· ⁀

14 4/25 ‖

* This code is not immediately obvious. Up to 5 km (5000 m) the figure on the chart indicates hundreds of metres. Above 5 km the figure represents kilometres plus 50, e.g. 60 indicates a visibility of 10 km.

The limitations of a single synoptic chart, however accurate and well presented, are that it cannot give us all the information available to the meteorologist when he was preparing his forecast; and it does not show trends. Further, for practical reasons it is unable to note all the local modifications to the weather caused by factors which will affect only the area concerned – such as fog on an estuary airfield. This is where we have to work out some possibilities or alternatives for ourselves.

Synoptic charts for general use give the weather situation at and near the surface only, and are called *surface* charts. They are sufficient for most needs, although if a long flight with few diversionary possibilities is planned, the more specialised *upper air* charts, which are available in most forecast offices, should be examined. The most useful ones are the 500 mb and 300 mb charts, corresponding very approximately to heights of 20,000 ft and 30,000 ft respectively. A synoptic chart may be drawn to show what is expected to happen, when it is called a *forecast* chart, or what did actually take place – an *actual* or *situation* chart. If possible both should be studied, but we should always check the dates and titles, since it is wasting time to try to make a forecast from information which is, itself, only a forecast (Figs. 3.3 and 3.4).

Interpreting the signs and symbols

The thin lines on synoptic charts, like map contour lines, are *Isobars*, and they connect all places having equal barometric pressure. On small scale weather maps, such as are printed in newspapers, isobars are usually drawn 4 or 8 millibars apart. The direction of the wind can be taken as more or less along the isobars, with generally some slant towards lower pressure. The strength of the wind is indicated by the distance the isobars are apart. If they are close together, this means strong winds as there is a considerable change in pressure over a relatively small distance, or in other words, a steep *pressure gradient*. If the isobars are far apart, there is a slack pressure gradient, and the winds will be light, even calm (Fig. 1.1). The symbol for wind is an arrow shaft lying parallel to the surface wind direction, and aiming *with* the wind. The feathers at the tail of the arrow (from where the wind has come) show the strength (Fig. 3.5). A single full length feather is 10 knots, and a half feather 5 knots; three and a half feathers, for example, indicate 35 knots.

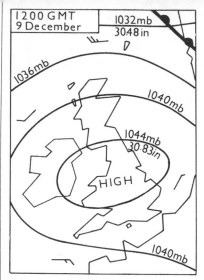

1032mb
3048in

1200 GMT
9 December

1036mb

1040mb

1044mb
30.83in

HIGH

1040mb

3.3 The weather map

A small anticyclone with a very high central pressure over the Midlands where it will be calm and probably hazy. Note the wind circulation; over a distance of some 500 miles there are only light winds – E in the South and W in the North. In winter radiation fog or frost will develop if the sky is clear though frequently the weather will be merely overcast and cold. Pressure is usually given in millibars, but on some charts inches of mercury may be used instead.

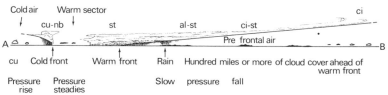

Cold air   Warm sector   ci

cu-nb   st   al-st   ci-st

A   Pre frontal air   B

cu   Cold front   Warm front   Rain   Hundred miles or more of cloud cover ahead of
                                                    warm front

Pressure   Pressure   Slow   pressure   fall
rise   steadies

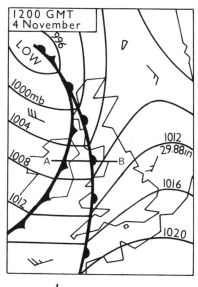

1200 GMT
4 November

LOW

996

1000mb

1004

1008   A   B

1012

1012
29.88in

1016

1020

3.4 The depression or Low

Forecast chart showing a depression crossing Britain in winter. North of Scotland the warm and cold fronts have occluded. Isobars are close together in the north showing strong winds. Note that wind direction is slightly across the isobars with an inclination towards the low-pressure centre. The weather at line A–B is shown above. You and your 65 h.p. pride and joy are too small to be visible.

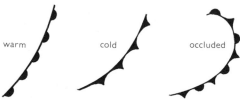

warm   cold   occluded

trailing
or
stationary

Shipping forecasts give the wind strength as a Force on the *Beaufort Scale* (page 224).

Although the isobars of a weather map show something quite close to the direction of the wind at a particular instant it must not be forgotten that all systems shown on the weather map are themselves wholly in motion. The path of an individual particle of air will depend both on its motion approximately along an isobar and on the movement of the system a whole. So in order to trace the path of a particle of air or the whole system, we should look at a series of weather maps extending over an appreciable period and work out the path by combining the two motions.

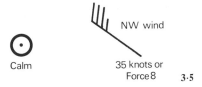

Calm   35 knots or
        Force 8   3.5

Front lines

The thick lines on the chart, decorated with semi-circles or spikes, which lie across the isobars mark the warm and cold fronts (Fig. 3.3). The passage of a front *always* means some change in the weather, but how much depends on whether the depression is active, or weakening, and how far we are away from the centre. Where a front line on the map has both symbols on it alternately, the fronts are occluded. Symbols are drawn on the advancing edge of front lines, but it should not be expected that fronts will necessarily move at anything like right angles to the lines. If a front is stationary, or slithering along its length, the symbols are drawn on both sides of the line; the warm or cold front signs being given according to the behaviour of the air in that area.

Some charts have arrows to show the estimated direction of movement of fronts or of a complete system, which may also be labelled with a letter for ease of identification. This is helpful when studying charts on succeeding days.

Pressure pattern variants

The synoptic charts (Figs. 3.3 and 3.4) showed a typical depression and anticyclone, but the pressure distribution is not always so

Pressure

The pressure tendencies at coastal stations are defined as follows:

STEADY	Change less than o·1 mb in 3 hours
RISING OR FALLING SLOWLY	Change o·1–1·5 mb in last 3 hours
RISING OR FALLING	Change 1·6–3·5 mb in last 3 hours
RISING OR FALLING QUICKLY	Change 3.6–6.0 mb in last 3 hours
RISING OR FALLING VERY RAPIDLY	Change more than 6·0 mb in last 3 hours

The speed of movement of pressure systems may be given in forecasts. The terms used are as follows:

SLOWLY	up to 15 knots
STEADILY	15–25 knots
RATHER QUICKLY	25–35 knots
RAPIDLY	35–45 knots
VERY RAPIDLY	over 45 knots

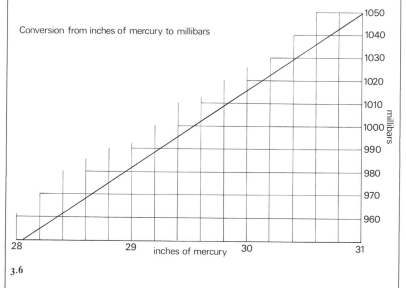

Conversion from inches of mercury to millibars

3.6

Highest and lowest recorded pressures

The highest pressure ever recorded (reduced to mean sea level) was 1076·2 mb at Irkutsk in Siberia, although in the winter of 1968/9 it is reputed to have been as high as 1080 in the same area. The lowest ever was 886·8 mb on a boat 400 miles E of the Philippines. The highest pressure recorded in Britain in the last 100 years was 1054·7 mb, and the lowest 925·5 mb.

tidily arranged. Occasionally a depression becomes blocked by a large anticyclone and is slowed down, only to be run into from behind by another Low. It is almost as though the anticyclone were a sort of sea wall with Lows, like waves, piling up against it. Low-pressure systems will also snake around the edge of a High with the fronts trailing along this sinuous line. When a front extends over a great north–south length, the prevailing air temperature may be very different towards the two ends; the southern tail end of a warm front may find air ahead of it warmer than itself, so will cease to ascend. Instead, it remains underneath and becomes, in fact, a weak cold front. It is this sort of situation which may cause a little kink or wave to occur on the front line, which sometimes develops into a new, secondary, depression.

A *trough* may be just a little tongue of low pressure or it may be a much more extensive line or belt, which develops when air finds itself warmer than the air ahead. This situation can arise, for example, in northern Europe in autumn when air coming in from the still warm Atlantic moves into the colder land environment. Because of the ascending nature of air in a trough, the weather will be cloudy, wet, and sometimes squally or thundery (Fig. 3.7).

3.7

A *Polar Low* is a small low-pressure cell which may develop in a fresh cold airstream. It may be little more than a local uplift of air within the cold dense flow, but it may quickly develop into a tiny fierce depression. It comes on the scene quickly and is often difficult to forecast.

A high-pressure *ridge* is a narrow region of high, or higher, pressure between areas of low pressure. It is vee-shaped with pressure highest along the centre line. A ridge gives improved, and often very fine, weather, but this does not last long at any point on the ground as it is being shifted along, usually fairly quickly, by the adjacent low pressure systems that caused it (Fig. 3.8).

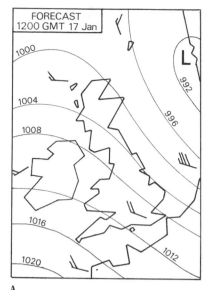

FORECAST
1200 GMT 17 Jan

1000
1004
1008
1016
1020
1012
992
996

L

A

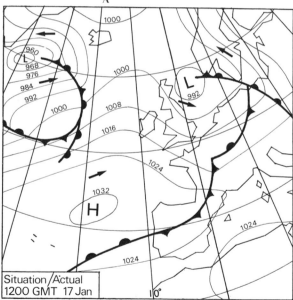

1000
960
968
976
984
992
1000
1000
1008
1016
1024
1032
1024
1024
1024
992

L

L

H

Situation / Actual
1200 GMT 17 Jan

0°

B

3.8 A ridge of high pressure

Especially in winter a ridge, however weak, comes as a pleasant interlude in long spells of wet, cold or windy frontal weather. Suddenly, if only for one day, it is flying for pleasure. The first chart shows a forecast ridge due to peak over central Britain during the afternoon. At the forecast time of noon, the wind will be WNW, and quite strong at 20 knots, although probably not gusty. Over most of the country, except where there is ascent of air forced up by hills, the sky will be blue or only partly cloudy. As the ridge moves across, the wind will slowly back westerly and then SW. With this backing the first effects of the next approaching depression will be felt. The wind may become more gusty, and high cloud will probably be visible in the western sky. By nightfall it is likely to be raining in Cornwall, and by next morning everywhere will be wet once more.

The two charts, forecast and actual, should be compared to see how the forecast worked out. The 600–700 miles that the next Low will have to travel to reach Britain is quite a normal 24 hour trip for an Atlantic depression.

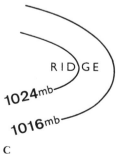

RIDGE
1024mb
1016mb

C

If pressure rises rapidly, the approaching High is more likely to be a small ridge than a developing anticyclone; this is usually heralded by a slower pressure rise.

Jet streams

A jet stream is a high-level flow at 30,000–40,000 ft producing strong westerly winds of 100 mph or more; 400 mph having been encountered by airliners in the region of Japan. The energy of the jet stream results from the fundamental and huge transfer of heat between the tropics and the poles. Along the top of this wavy and sometimes discontinuous region which encircles the earth in both hemispheres vortices are created. Their width is generally between 60 and 80 miles and the air in the core is smooth with considerable turbulence around the edges. The effect of the rotation of the vortex is to pull down the warm air on the tropical side, and push up the cold air on the polar side; as would be expected, this action will result in cloud and precipitation along and under the cold side. It is the large temperature changes across the vortex stream which cause the strong winds (Fig. 3.9).

Jet streams occur in many different parts of the world; the start of the monsoons in India, for example, coincides with the jump of the summer jet stream which flows to the north of the Himalayas, to the south of those mountains in the autumn.

Although jet streams are used deliberately by airline captains to save time and fuel, their main interest for us lies in what their

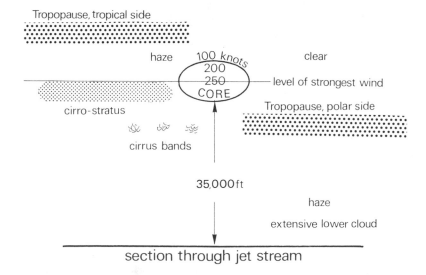

section through jet stream

visible cloud can tell us about the weather to come. As with most other aspects of weather, quite recognisable patterns exist once we get to know them, so it is not surprising to discover that there are identifiable jet streams, even with names. The one that probably affects us most in Britain is the maritime Arctic-front jet. This flows out from the eastern seaboard of the United States at a height of about 32,000 ft, and has a core temperature of about $-35°C$. It is stronger in the winter due to the steeper temperature gradient between the cold North, and the air from the tropics. This stream may continue across the Atlantic and more or less over Britain or it may split, one half swinging north over Iceland and Scandinavia and the other going further south over North Africa. The importance of this to our weather is that a lot of depressions track along the line of the jet stream. If the jet stream is split the Lows will tend to follow each other away to the north, and western Europe will grow a large and persistent High, with the weather in Britain being warmed by air from the south. If, however, the jet continues on a generally easterly heading, depressions may continue to trail over Britain for a week or more. The existence of a jet stream can be recognised by long streamers of cirrus moving at easily discernible speed across the sky. Usually the cirrus will follow the typical westerly flow, although it is possible for smaller N and S jet streams to occur over Europe. So if we see jet stream cirrus streaming across the sky from the west we are warned (Fig. 3.10).

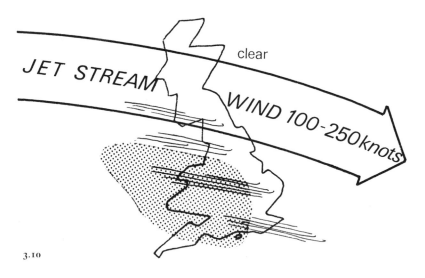

3.10

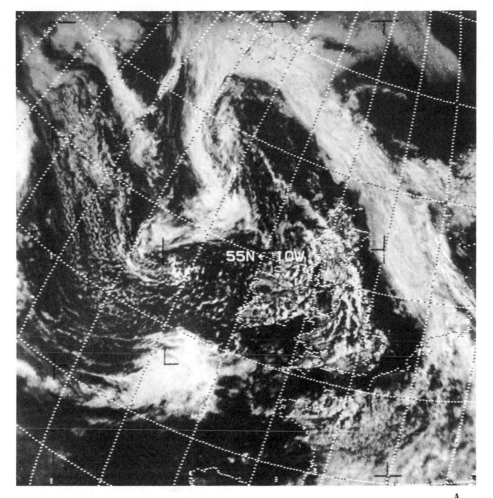

A

3.11 Satellite Essa 1400 hrs

9 May 1972

A minor trough of low pressure within a homogenous airmass is lying the length of Britain (Fig. 3.11B) and in the uplifted air is producing a line of thunderstorms which can be discerned on the satellite photograph (Fig. 3.11A). In the biggest of these (over the S Midlands) a glider flown by Mike Field climbed to 27,000 ft eventually coming out of the storm on the W side. Here he found wave lift and climbed to over 40,000 ft, the highest a glider has ever reached in Britain. The wave was caused by

—continued

a strong upper air flow rising up over the line of cu-nb which had grown up to the tropopause in its path, and which was still moving slower than the wind at that height. The chart shows the isobars at the surface with the wind strength and direction at 30,000 ft superimposed. The 100 knot wind in the SW approaches was produced by a jet stream at this height aiming for Brittany.

The satellite photograph shows the occluded front lying to the NE of Britain and the small low centre to the NW. The absence of convection clouds in the sea areas around Britain is marked.

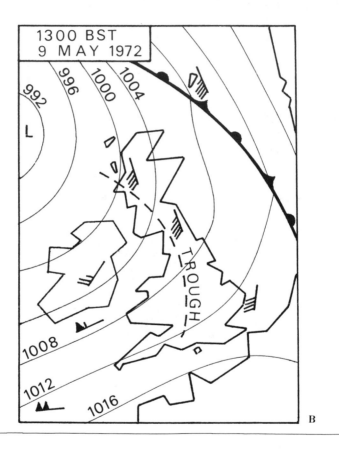

Air in limbo

Unfortunately for both the forecaster and the pilot, weather is not made up only of neat and tidy pressure systems. Between the more recognisable and vigorous circulations the air is a rubbish dump of time expired weather. In general this waste land consists of small variations in temperature, pressure and humidity which may produce indeterminate patches of cloud, perhaps slight precipitation, or even a clear sky. Whatever the conditions they are unlikely to last long as they will quickly be invaded or consumed by new or stronger depressions and anticyclones. In northern Europe quiet, nondescript weather with pressure on the low side should be regarded as both temporary and suspect.

4   Analysing the weather map

With the homework on basic weather structure done, and the signs and symbols learnt – or at least to hand – we can get on to more practical, and fascinating, aspects of met. This chapter is really detective work on weather maps, to see in detail just how much information can be got out of them. In succeeding chapters we will get out and look at the weather visually so that we can link the paper chart to the real live sky.

Both *The Times* and the *Daily Telegraph* give good maps each day, as does *The Guardian*, and for weekend fliers the *Sunday Express*. *The Times* prints a forecast map of the North Atlantic area, and a more detailed forecast map of the UK for noon on the day of issue of the paper. These papers give additional summary information such as world temperatures, and skiing or sea passage information, as well as a written prognosis. This extra information may be more useful than at first appears.

The synoptic charts in Fig. 4.1 are from *The Times*, both maps being, as already said, forecasts for noon the same day, and therefore giving the same information. The chart of the Atlantic (A) shows a deep depression of 960 mb at the centre located between Scotland and Iceland, the influence of which is affecting all of the British Isles. The cold front has already crossed the country, probably during the night, and is now over the Low Countries. Since it almost certainly gave rain the ground will be wet. The wind over Britain is SW, light in the south-east of the country and, with the isobars closer together, increasingly strong towards the north. It will be moist from its long passage over the sea, and cold because it is sweeping down and round from the Arctic Circle. In such a situation it would be useful to check the temperature and conditions in Reykjavik, Iceland, in the World Weather column, to see how cold the air is there.

Since the centre of a depression is effectively a great hump of expanding and cooling air it will almost certainly be filled with cloud and rain. Whether or not this cloud extends solidly as far as Scotland, more cloud will in any case be produced where the moist Atlantic air flows over the land; this cold, damp, airstream will be

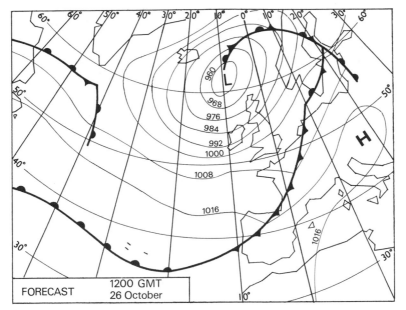

FORECAST | 1200 GMT 26 October

A

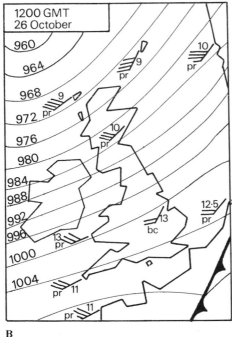

1200 GMT 26 October

B

4.1 Weather maps redrawn from *The Times*

A deep depression near N. Scotland will move slowly NE. All districts will have showers and sunny intervals. Over much of Northern Ireland and Scotland the showers will be frequent and sometimes heavy, with a risk of hail and thunder. Some heavy ones are also likely over England and Wales, especially in W districts. S to SW winds will be strong or gale force in many districts. Temperatures generally near normal for late October. Arrows show wind direction. Figures outside the circles show temperature. Weather Symbols: b, blue sky; bc, half clouded; c, cloudy; o, overcast; f, fog; m, mist; d, drizzle; r, rain; h, hail; s, snow; th, thunderstorm; p, showers; z, haze.

pushed up by the high ground and cooled. In addition, should the land be warmer than the sea, air flowing over the ground will be warmed and will expand, rise, and be cooled adiabatically. All this will add up to a considerable amount of cloud, maybe clear enough to fly underneath, but quite possibly not. Much of it will be thick enough to give rain, or at least showers, over the west coast hills. In the higher part of the cloud the air will be cold enough to produce snow, although this may not reach the ground; over the Scottish hills there may be hail and even thunder.

If we now look at the map of the British Isles we find that some homework has already been done. The arrows confirm that the wind will be SW, but they also give the speed – 40 mph in Scotland, 30 mph along the western coasts, and 20 mph in the SE. These winds are all a little stronger than might appear from looking only at the Atlantic map.

Although the wind in the SE of Britain will have been slowed down more than anywhere else by its passage over the land, the reduction in speed will be most apparent on the ground; it should not be expected to slow down much at flying levels.

The temperature is fairly consistently around 10°C except in the extreme north. Again, conditions are better in the SE where the temperature is up to 12°C.

A very general idea of the cloud amount, but not of the type, can be obtained from the small letters by each wind arrow. They show that cloud will be extensive enough to cause precipitation on all western coasts where the moist air first hits the land, except where it is in the lee of Ireland, which has already done the job. This we had not taken into account in our private deduction. Once again the SE comes off best with half-clouded. This might mean little more than intermittent breaks, but it should be enough to see something of the ground all the time.

Let us now see what further information can be deduced from the chart, even though it is not stated. Because the air is obviously very moist it should be expected that cloud base will be low; it is likely to be very low in precipitation, where visibility will also be poor. Routes should be planned to avoid high ground. In this sort of airstream the temperature of the air will be dropping with height at about 3°C per 1000 ft – the dry adiabatic rate – so freezing level could be as low as 3000–4000 ft, with icing in cloud. It is therefore not an ideal day for trying to keep a date in a small aeroplane flying northwestwards from London.

Tomorrow? The well established High in central Europe will ensure that the depression is deflected to the north of it, towards Scandinavia. There is another depression from south of Greenland which is developing and will probably move eastwards quite fast; nevertheless, there is the possibility of some clearer weather with only moderate to fresh westerly winds for at least some hours before the influence of this approaching depression is felt.

The next chart (Fig. 4.2) is from the *Daily Telegraph*. It gives a situation report on the actual weather at noon on 26 February and a forecast for noon on 27 February (the day of issue). We are at an advantage here since we can compare the weather we actually saw with the map. The Atlantic chart (A) shows an anticyclone centred over Scotland with light winds over most of the country; the air is almost calm in the north, with winds strongest along the south coast. The central pressure is quite high – 1032 mb. We saw, ourselves, that this day was fine and dry, although with some cloudiness. Being February, and with the air over Britain coming from east Europe, it was cold. In the north of the British Isles it was less cold because the air had been for a longer time away from the chilly continental influence in its passage over the eastern Atlantic. Here it would have been over warmer sea for some time, since the winds are very light, and adjacent to the warm southerly flow from the Azores. In such a weather situation it would be useful to look at the temperatures for, say, Edinburgh, Copenhagen, and Berlin, in the World Weather column.

Studying this situation Atlantic chart, and without looking at the small forecast chart (B), we should try to work out how the weather will change, and then see if we have come to the same conclusion as the met man. There is little doubt that the High pressure will remain established over Britain for a time, since it is relatively high over the whole of north Europe. Such a large anticyclone will tend to block the approach of the depression out in the Atlantic. This Low, however, is also large and powerful, with a central pressure down to 968 mb, so it will not just wither away. There is already the strong southerly flow in mid Atlantic, so the probable course of the Low will be to slide up the isobars in the direction of Iceland and towards Finland. This could result in some displacement of the centre of the British High towards the east or north-east, which would have the effect from bringing warmer air nearer the western

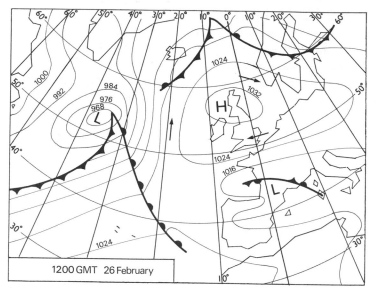

A

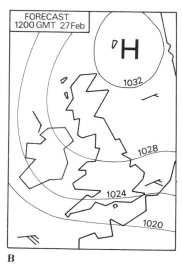

B

4.2 Weather maps redrawn from the *Daily Telegraph*

An anticyclone of 1032 mb is centred over Scotland. Winds in southern Britain are easterly, and cold, having come from eastern Europe. The temperatures in Ireland and Scotland are higher, the air originating from the warmer Atlantic. The winds everywhere are light. There is probably extensive rain over France from the occluded, and almost stationary, front. The forecast chart shows the High centre drifting NE and contracting so that pressure is lowering in the SW approaches.

parts of the country.

Looking now at the small forecast chart (B) we see that the expected movement of the high centre has occurred, but in a more northerly direction. Temperatures are unlikely to change much, because the air over Britain, except in the extreme west, has still come from the cold continent.

Weather maps in sequence

Whenever possible, and certainly before any important flight, we should study charts over a period of consecutive days, as this will give a clearer picture of how the pressure pattern is changing. We should note the central pressure, and the steepness or otherwise of the pressure gradient – the change of pressure across the isobars – and the rate of development or decay in the system. The sequence of charts will also make apparent the direction and rate of movement of the systems, or whether they are becoming more or less stationary. With a sort of motion picture in our mind, it is easier to visualise what the weather might do, and how the air will behave on its travels. Figs. 4.3 show the weather over six infamous February days, and gives a picture of the life and eventual collapse of an extremely unpleasant depression. The weather was so bad between Scotland and Iceland that trawlers overturned and sank with heavy loss of life due to the sheer weight of ice on their superstructure.

The first chart (4 February) shows a deep depression of 960 mb centred on Iceland. North of that country the winds are easterly, sweeping down at almost gale force from the Arctic to become cold westerlies over Britain. The depression is already partly occluded, but still vigorous. 5 February shows the centre as having moved south, with the almost entirely occluded front over NW Europe. The Low is weakening slightly with the pressure at the centre having risen to 984 mb, but the winds are still arctic born, thrusting along at a good speed over the cold winter sea. 6 February shows the low centre still in much the same place because further eastward movement is being blocked by a massive anticyclone – a central pressure of 1040 mb was reported from the USSR. In stagnating, the Low centre has broken down into 3 cells, although still with the central pressure below 1000 mb. Winds are now less strong over Britain, and more variable in direction, but still bearing some squalls. By 7 February the frustated Low, still blocked by the Siberian High,

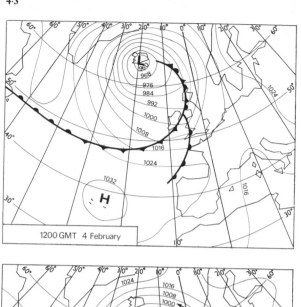

1200 GMT 4 February

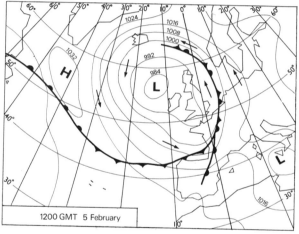

1200 GMT 5 February

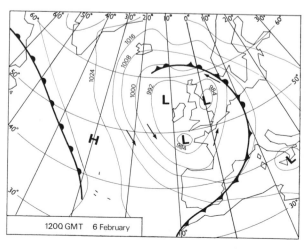

1200 GMT 6 February

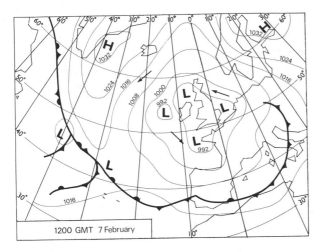

1200 GMT 8 February

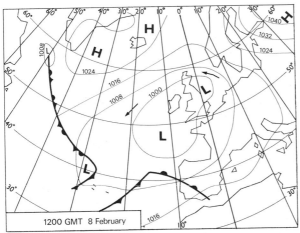

1200 GMT 7 February

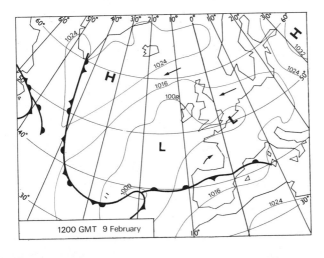

1200 GMT 9 February

is weakening and filling, and being further pressed by a developing High in the NW Atlantic. It is just dying on its feet. During 8–9 February it finally expired. Winds have become lighter in the slack pressure gradient, and the air which is now arriving from a gentle meander over southern Europe is not so cold. And the outlook? Pressure everywhere in the Atlantic is fairly uniform, the High off Greenland having also weakened. There is, therefore, little to stop the newly developing Low off the eastern United States seaboard from charging across the Atlantic. What happens to it on arrival will depend mainly on the state and influence of the Siberian High, which is still lurking up in the corner of the chart.

In winter cooling of the air may produce rain, or snow. Flying in rain is merely dreary, but flying in snow can quickly become hazardous. It is helpful, therefore, to be able to know when and where rain is likely to change into snow, and vice versa. In the first map of Fig. 4.4 there is a wave depression lying over southern Britain. As would be expected this line of expanding air trailing in from the Atlantic and being raised and cooled by the land, is producing rain. It is the sort of rain that may continue for several days, sometimes drizzly, then heavy, then nothing, and then more rain. Unpleasant weather, but nothing to be alarmed about.

The next day, 23 February, the anticyclone over Spain moved west and weakened and the low-pressure area moved SE into Europe. The winds, although generally light, were now faced with the Alpine massif, and forced up high enough to produce not only rain, but snow. A low-pressure cell developing SE of the Alps contributed to the expanding and cooling behaviour of the air over the whole of the Alpine region, so the snow increased, with blizzards in the mountains lasting for several days. Ski resorts were cut off, many people were killed by avalanches, and air traffic was completely halted. If we can foresee this sort of situation it can make the difference between deciding to start home early, or sitting out an enforced stay of maybe a week at some inconvenient airfield.

The risk of sudden or unexpected snow in the British winter is greatest when there is a dry cold northerly airstream, having perhaps a pressure of 1000–1010 mb, blowing over the UK. Small low-pressure cells may develop with little or no warning in this polar air and are swept along in the main flow. When the lower pressure air reaches the higher land its pressure is reduced further and it is rapidly cooled. Cloud develops, producing snow, sometimes in

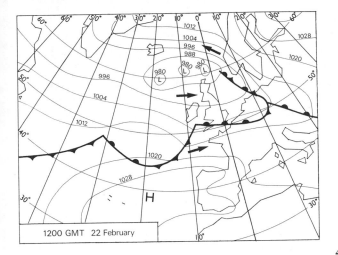

1200 GMT 22 February

4.4 Snow over the Alps

22 February. A wave depression
running along the English
Channel and the south coast is
giving prolonged rain over these
areas. On 23 February the high
pressure over Spain weakened
and the wave depression chain
broke, parts of it moving south.
As the cold air approached the
Alps it was massively raised, and
cooled further. The rain became
snow, which was so heavy that
blizzard conditions lasted for
several days. There were several
avalanches with loss of life.

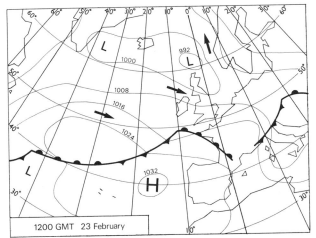

1200 GMT 23 February

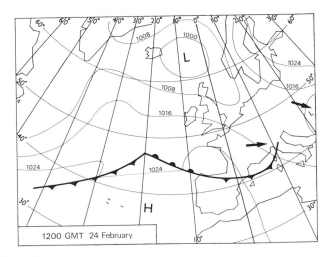

1200 GMT 24 February

unreasonably heavy amounts, on a day that was expected to be fine, though perhaps bracing. Often the snow falls only along the track of the small low-pressure cell, but if there is snow falling more widely, it will be heaviest along this track. Further inland the snow will probably die away as the air is progressively dried out – unless there is higher ground ahead. This is a situation in which the direction of the wind is very important since the snow may occur only along a fairly narrow belt, or where the air is first raised over the land. Elsewhere the weather may well remain entirely reasonable for ordinary flying.

Fig. 4.5 shows the area over which there was extremely heavy snow, some of blizzard intensity, in relation to the wind direction. When flying in the British winter on a day with cold dry air tracking down from the Arctic, we should expect tiny, unforecast, low-pressure cells to develop, and have an alternative plan in mind, particularly if flying anywhere near north or east coasts.

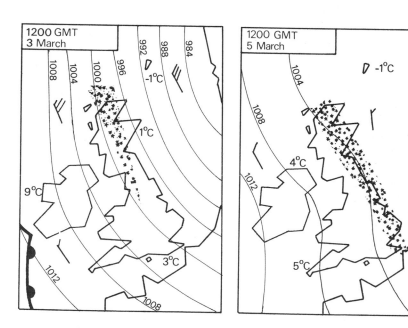

4.5 Snow over Britain

If a small low-pressure cell develops in a cold northerly airstream, snow-bearing cloud will be produced when it moves up over the land. The areas of land first reached by the cold wind will get the snow, which may be of blizzard proportions.

Patterns of weather

In most parts of the world, including the British Isles, certain patterns of weather will recur, sometimes quite frequently. Even in Iceland, where the weather changes more rapidly than nearly anywhere else, a particular synoptic situation will repeat itself – almost giving the feeling that we have seen all this before. Regular inspection of weather maps over a long period helps us to recognise these repeat patterns, and makes it easier to forecast what is going to happen next.

One such situation, mainly due to Britain's position off the coast of a large continental land mass, occurs when there is a large anticyclone over mainland Europe. If the high-pressure area becomes sufficiently well established to last for several days, depressions approaching from the Atlantic may be blocked, as we saw in Fig. 4.4, or distorted or diverted. The extent to which a Low develops a circular shape, or becomes a sinuous wave depression, or is otherwise modified, depends on the location and movement of the High, and the relative temperatures and pressures of the two air masses. Often the fronts will slide along and around the western edge of the High over much the same line on the ground, giving continuing foul weather locally while the rest of the country has it fine. These trailing fronts are a real headache to the forecaster, as the bad weather is lying along a relatively narrow strip moving more or less along its own length. The difficulty is not so much to predict the line of bad weather, but to anticipate any sideways shifts that it may make which will bring the cloud and rain to places counting on a fine day. As the anticyclone weakens the fronts will be able to intrude further, and the wave snake will break.

The situation chart for 27 March (Fig. 4.6A) gives a vast High over the whole of Europe. Although it is only early spring the air over the continent is not cold. This is largely because there is almost no wind and the circulation of the High, such as it is, is using more Mediterranean air than Siberian. The High extends over southern Britain and there is a trailing front associated with a depression centred immediately west of Iceland. The trailing front is kept at bay by the high pressure, which is also having the effect of 'squeezing' the flow between Britain and Iceland, and producing strong winds. The forecast map for 28 March indicates that the High is either moving south or declining, since the 1032 mb line has

4.6 Patterns of weather

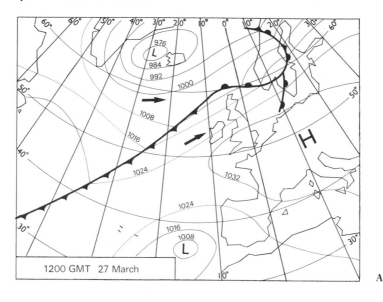

A

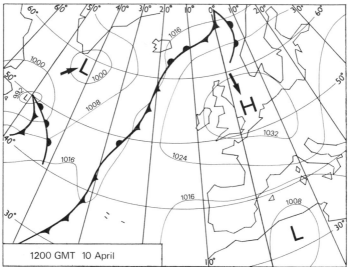

B

An anticyclone over Europe often gives a similar pattern of weather which affects the British Isles considerably.

Trailing fronts develop and move along their length, bringing bad weather to quite small areas for several days.

Illustrated are the situations on 27 March, 10 April, 29 29 September, and 31 October.

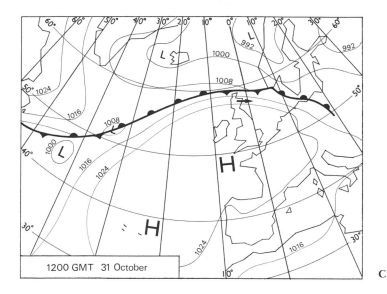

1200 GMT 31 October

C

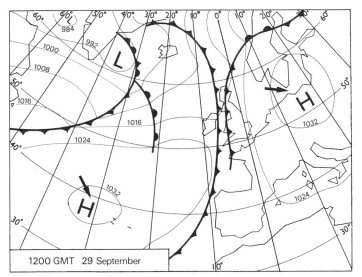

1200 GMT 29 September

D

On 31 October Scotland has warm anticyclonic weather, but a slight decline in the high pressure will permit the frontal rain to penetrate south and bring a marked temperature drop.

On 29 September (D), the trailing fronts are lying over west Britain while the east stays in sunshine. The cold fronts are piling up like waves on a shore, as the anticyclone prevents further eastward penetration.

Analysing the weather map / 57

retracted from the middle of Britain to the English Channel. This has allowed a southerly movement of the front line.

The forecast map for 10 April (Fig. 4.6B) shows a similar situation except that the High centre is further NW over the North Sea. Much of the air in this anticyclone has been influenced by a cold NE Europe, so temperatures everywhere are lower than earlier in the year on 27 March. For the time being, because the pressure gradient is fairly slack most of the way across the Atlantic, the synoptic picture is unlikely to change much. Generally the weather will stay fine over Britain although some rain may occur where the air from the sea first arrives over land, such as on the English east coast.

The situation map for 31 October (Fig. 4.6C) shows the Azores anticyclone extending over Britain and west Europe. The front-line is lying almost W–E to the north of Scotland, moving along fast in the fresh winds. The warmest weather in Britain was in Scotland itself, but only a very slight shift or lowering of pressure in the anticyclone will bring the front, with its cold air and rain, quite quickly over the land.

On 29 September (Fig. 4.6D) there is a well established anticyclone of 1032 mb centred over Poland. The winds are slack, and the air influencing France and eastern Britain will have come from the Mediterranean and be warm. The maps show a front of north Atlantic air stationary on the fringe of the High, with another wave of cooler air piling up behind it. This is giving a marked change of weather across the country. The east is basking in hot sunshine with a temperature of 20–25°C, whereas less than 100 miles west there is continuing cloud and rain with temperatures of only 15°C. Again, a slight shift of the High or its decline, will allow *rapid* penetration of the fronts across the country.

It will be apparent from all this that although weather is continuously changing, there is often a pattern, and always a reason for what happens. When examining a chart it is not only necessary to obtain a clear picture of what the situation actually is, but what reasons exist, or are developing, which will cause it to change. It is also not enough to look at a front and decide how it will move locally. The big picture has to be taken into account. When flying in Britain or NW Europe, *every* depression and anticyclone on a chart should be examined to discover which ones will dominate the future situation, growing or moving fast, and which ones are declining, and so will slow down or diminish. In this way a clearer idea of the range of possible changes to the weather can be obtained – is the front likely to come in faster, or slow down, or slither along its length? Is the high pressure building so that the weather will remain settled for a day or so? What is likely to happen to that little Low off Greenland? How far into Europe is the high pressure in the eastern Mediterranean likely to penetrate?

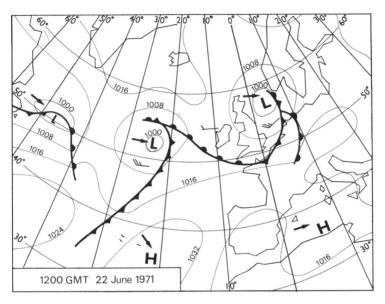

1200 GMT 22 June 1971

4.7 When a train of depressions develops it may bring a series of dry days with rain at nights, or vice versa, for sometimes as long as a week. Here London would have a fine morning and Cornwall a wet evening. After the mid Atlantic Low has followed through the developing Low off the Eastern US seaboard will be in a position to continue the series.

The second question that should always be asked is 'Where has the air come from, and, if the weather will be as forecast, where will new air be from?' Fig. 4.8A shows a depression in the Atlantic and a High over central Europe. On 14 October the air over southern Britain was moving up gently from the Mediterranean. The temperature in London was 20°C, 68°F, hot for October. As the Low centre moved NE to Norway, never coming nearer to the British Isles than 500 miles, its circulation reversed the flow. Now, only 5 days later (Fig. 4.8B), instead of drifting up from the Med. the

4.8 Where has the air come from?

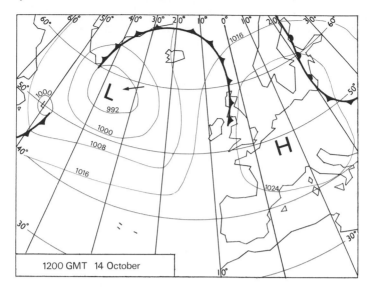

1200 GMT   14 October

A

A Southern England basking in warmth from the Mediterranean.
London temperatures 20°C, 68°F.

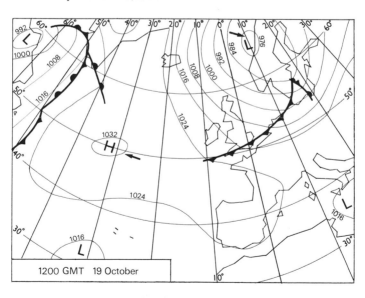

1200 GMT   19 October

B

B Only 5 days later: the Low is pulling air from the Arctic over
Britain. London temp. 14°C, 57°F, and feeling much colder in the
strong winds. In Iceland, there was real winter weather.

air over Britain was whistling down from the Arctic; temperatures dropped to 14°C, 57°F, and very strong winds made it feel even colder. Iceland was having a premature sample of real winter (Fig. 4.9). The cold front over southern England was created not by the usual birth processes of an Atlantic depression, but simply by the cold northerly blast sweeping up the old warmer air in its path. It created cloud and showers, rather than a line of solid bad weather.

4.9 Iceland weather. 19 October

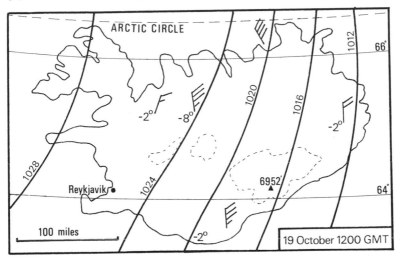

In Britain it suddenly became a bit chilly and draughty on 19 October, but in Iceland temperatures were below freezing over the entire country. It snowed, in some places heavily, and the winds over the central area of the country were blowing at 50 mph – blizzard conditions. In Reykjavik, the capital, where there is some shelter to the lee of the mountains, the winds were less unpleasant at a mere 15 mph.

Unexpected snow

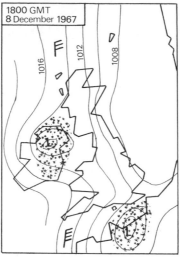

1800 GMT
8 December 1967

4.10

On 8 December 1967 snow fell unexpectedly on the south coast at Brighton to the depth of almost a foot, to the consternation of almost everyone in the area.

It had been cold all over Britain with a N wind for some bracing days; the ground level temperature having remained below freezing, but the sea still relatively warm. This situation sparked off a small fierce depression within the northerly airstream to the NW of Scotland. It travelled fast and arrived over Northern Ireland on the evening of 7 December and scattered it with snow; travelling at 30 knots it left a white trail across Wales and the West Midlands, and the forecast warned of slight snow for southern England.

By 0900 hrs on 8 December the centre of the Low was over the Isle of Wight and snow was falling all over the southern counties except for a narrow strip along the south coast – not an unusual state of affairs as people who moved there for the good weather will tell you.

Inland, traffic was slowed by snow lying in a temperature of $-3°C$, which is also the temperature below which salt ceases to help keep roads clear. Suddenly, at 1000 hrs, it started snowing heavily along the south coast and continued to do so for 8 hours, although nowhere else had snow lasting for more than 3 hours.

The probable reason is that as the cold air in the Low swept out over the warm Channel waters the Low intensified, and was also influenced by another small depression over N. France. Convection developed producing thunderstorms over the Channel. These grew, and the whole ascending air mass gave the heavy snow over the south coast and the sea.

Metmaps

Valuable practice in weather map interpretation can be obtained by making them up yourself. It is possible to do this by plotting the information obtained from the radio shipping forecasts on to specially prepared sheets, called Metmaps (Fig. 4.11). When the latest data from the radio has been plotted, it may be practicable to build in additional information from the earlier newspaper charts to obtain a somewhat fuller picture. Metmaps come from the Royal Meteorological Society, Cromwell House, Bracknell, Berks, with instructions on how to use them, and additional information and conversion tables.

After a while it is possible to derive from synoptic charts and forecasts a fairly good idea of how and why the weather will change, or at least what the alternatives are. But even thorough academic study is only part of the answer, since we have to relate it to what is actually happening to the weather outside, and then feed this information back into further study of the chart. The next step, therefore, is to learn to recognise in the appearance of the sky and the clouds the weather that is symbolised on paper.

How to use Metmaps

Metmaps consist of a chart of the British Isles and surrounding sea areas as far as Iceland and Portugal. This is supplied together with a blank forecast table on which to plot the BBC Sea Area Forecasts. In addition the Metmap includes a great deal of useful information including times at which forecasts are put out over the radio. The Metmap is used to plot the radio information so that a synoptic chart can be drawn. This chart will give the latest weather situation but only over the somewhat limited area of the British Isles (as can be seen from the lines on the example Fig. 4.11). This may be all that is needed, but often we want a wider picture. With practice the information from the older, morning newspaper chart can be incorporated, as has been done on the example with pecked lines. Obviously care is needed in doing this as some of the earlier data may be no longer relevant. The latest radio information should be used as a datum and the additional information just sketched in to give a fuller picture. Like any useful skill, learning to make a good synoptic chart quickly and accurately takes a little time. To begin with it seems impossible to write down the information in the time allowed by the announcer, so a tape recorder will be useful.

—sample chart over

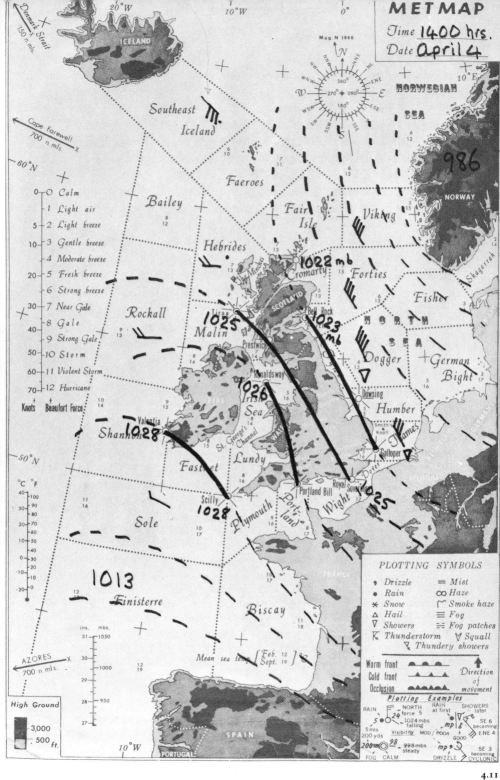

4.II

Part 2　Weather recognised

5 Reading the weather

We can only get full value from synoptic charts when we are able to visualise the actual real-life weather indicated by all the lines and figures. It is only half the story being told, or seeing on the chart, that an occluded front will lie 100 miles to the west of London, Geneva, or Memphis, Tennessee, if we do not appreciate what this will mean when flying there in a small aeroplane. We need to be able to recognise from the appearance of the sky what is happening, and whether the forecast is working out or not. Met men are not often wrong in determining the structure of the weather, but estimates of timing sometimes go awry. This is because small, and seemingly insignificant, changes in the pressure pattern at high levels or hundreds of miles away, or even right where we now are, will modify the big picture so that a front arrives early, or hangs back, or the rain or the clearance fail to materialise when expected. The meteorologist took all the information into account and produced a forecast based on what was available to him at that time. But the weather went on with its interminable changing, so from that minute on the data became increasingly out of date. When we meet the weather that we heard so much about it is no longer the same, but we can *see* what it is like *now*. Every change that takes place is there in the sky waiting to be judged, because the clouds reflect faithfully the behaviour of the air that created them. *As the weather changes so do the clouds.*

Sometimes the 'visual forecast' range can be large, as for example, when the appearance of high level cirrus cloud gives us a direct link with a front perhaps 200 miles away. At other times it may be small and local, but important, like a single unexpected rag of cloud on a hillside. The amount of information handed out visually, and gratis, is enormous and it can be read like a book. But I mean a book and not a poster; a glance is not enough.

To understand the meaning of changes in the look of the sky we have to know the direction from which the weather is coming, and notice small alterations in that direction. Some people find this easy because they orientate themselves instinctively; they always seem to know where North is or from which direction the wind is blowing;

Red dust: where did it come from?

When getting a forecast we should try to discover where the air has come from which is dominating the situation, because its characteristics will determine the amount of cloudiness, visibility and temperature.

Sometimes the source of the air is of particular interest, as when coloured dust or rain falls. The most remarkable recorded occasion was back in 1755 when 9 in. of red rain dropped on Locarno one October night – which included an inch of red mud. This of course dried into a

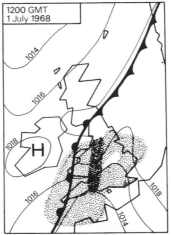

5.1

horrible dust. The area covered was about 360 square miles. High in the Alps 9 ft of red snow fell at the same time.

On 1 July 1968 a total weight of about 1 million tons of dust fell over England and Wales – although this only gave an average depth of a mere micrometre. It showed on satellite pictures and so its travels could be analysed. On 29 June, two days earlier, the dust was above the Mediterranean between 30°–40° N and 10°–17° W. On 30 June it arrived over England in a layer between 10,000 and 17,000 ft. The dust provided a good source of condensation nuclei and alto-cumulus-castellanus (see p. 79) clouds soon developed within the dust layer. It was the rain that subsequently fell from these high clouds which deposited the dust.

Further investigation indicated that the dust left the vicinity of the Ahaggar Mountains in southern Algeria in a big thunderstorm on the afternoon of 26 June. It was grey, yellow and brown, composed of sand, quartz, and a little clay.

Dust almost certainly drifts northwards from Africa quite often, but does not reach Britain due to having fallen out in rain on the way, or it is slight and goes unnoticed. The Sahara desert exports a good deal of dust, some of it travelling quite regularly to Barbados on the Trade Winds.

others find difficulty in locating themselves without the aid of a compass. They are at a disadvantage because orientation is essential for visual forecasting. If necessary, we should practise working out wind changes in relation to a fixed known reference point such as an office window, or the sun.

The cloudy sky

To be able to recognise the significance of any particular cloud or cloudy sky, we need to know why clouds differ in shape, from which ones rain or snow will fall, and how long they will last – because clouds have a life-cycle just like plants or humans.

We know that cloud is produced by the condensation of water vapour in the air, and that for any given temperature air is able to carry only a certain amount of this moisture; the quantity being greater in warm air than in cold (page 19). When air is sufficiently cooled by expansion, or from contact with a colder air or surface, some water vapour condenses into minute droplets – cloud. *The shape and appearance of this cloud is determined by the manner of cooling, the rate at which cooling takes place, and the state of the surrounding air.* Generally cloud in layers or sheets is caused by gentle rising of air over a large area, cumulus clouds are the result of numerous small scale upcurrents, and hard-edged lens-shaped clouds are formed by wave motion in the atmosphere.

Fog

Fog is cloud in contact with the ground, and hill fog is cloud covering only the higher ground. Radiation fog is cloud that forms in low lying places on clear calm evenings when the warmth that has been gained by the surface during the day radiates out to space. The surface cools rapidly, cooling the air in contact with it enough to cause condensation. If this is only slight dew will form, but if it occurs throughout a thicker undisturbed layer, as in a sheltered valley, the mist or fog which results may be 50 ft or more thick.

Rain

For cloud to produce rain the minute cloud droplets have to increase in size until they are big enough to fall under their own weight. Generally ice crystals have to be present in the clouds to start the process. They accumulate water by collision and coalescence, sinking

It was only a wisp of fog . . .

We were flying Queen Bees (Tiger Moths to be used as targets), and being a new ferry pilot I was told to stay on the heels of my senior. It was a beautiful day and the Solway Firth was brilliantly clear except for rags and patches of sea fog, their tops shining gold in the sun. My leader was flying at 600 ft and furrowed through the top of a wisp whose flurries raced past my wings. A mile ahead the lead Bee vanished into another patch; I followed into the sudden greyness and then we were out again into the sun, the green sea clear beneath.

About a mile ahead was another chunk of fog, bigger but still looking shiny and innocent. My leader held on and we rushed into the face of it, although I would rather have skimmed across its top, feeling the speed. We didn't come out. Instead it grew darker so I closed up, terrified of losing my leader – which of course, promptly happened.

Alone, I fixed on the instruments, concentrating my whole experience of 4 hours in a Link and 15 minutes for real, trying not to cheat; also I was starting from no known attitude since my last reference had only been another aeroplane. Needless to say the Bee wound itself up rapidly into a shrieking spiral. With only 600 ft there was not much time to sort things out, and before I was anything like back on an even keel I gave way to the impulse to look out instead of only at the instruments. Ahead of the nose were just grey waves – in plan view. I shall never know what I instinctively did but the poor little Bee lurched level with just feet to spare. Shaking and horrified, and in a grey cave 50 ft deep with stalactites of fog connecting with the sullen sea, I found that I was fortuitously on a compass course taking me away from the land. I let it ride without caring while I settled down, hoping to come out of the side of the fog, but it didn't clear. It was hard to think tearing along in this shallow gloomy cave with visibility less than 500 yards, but nothing would have induced me at that moment to try to climb up through the fog – even if there was only 600 ft of it. So I decided to turn on to the reciprocal of the compass bearing which should, I hoped, bring me back to the cliffs at an angle – and to see them in time. I watched for the gloom to darken ahead and at about the same time as the cliffs materialised I saw an isolated lump of rock growing up into the fog ceiling. With the cliffs in view life improved immediately and I flew along them just above the water and to landward of the rock. The fog lightened and with a bright flash I was out in the welcome clear.

When next I flew that way there was no fog, and as I passed I looked' down for my friendly cliffs and the rock that soared into the fog. I saw both – and there was a lighthouse on top of the rock.

—Author (1941)

through the cloud. When they have become heavy enough they fall through the bottom of the cloud as rain. This usually happens only when the cloud has grown to some 8000 ft or more in height, with the tops above freezing level. Water droplets in cloud do not automatically freeze when the temperature falls to 0°C, but mostly remain in a supercooled state. The temperature has to fall to $-40°C$ before all water droplets freeze, but between freezing level and the height at which the temperature drops to $-40°C$ there will be an increasing number of ice crystals among the supercooled droplets. As the cloud grows taller there will be further opportunity for the falling ice crystals to accrete moisture. The colliding process is increased in turbulent cloud, and this is one reason why rough April shower clouds produce more rain than the equivalent sort of cloud in the quieter air of Autumn. It is also possible for rain to fall from clouds which are less thick, and in which·there are no ice crystals. Most such rain comes from large cumulus over tropical oceans, or cloud near coasts. Instead of ice crystals it is large salt nuclei that accumulate moisture until the droplets are heavy enough to fall out. Rain caused in this way is unlikely to give more than a shower.

Hail and snow

Unless the temperature in the surface levels is extremely low, as it is in the polar regions, all lower cloud will be composed of water droplets, and rain will fall from them. If, however, the cloud grows high enough to form ice crystals, and if these crystals accrete and freeze on to themselves sufficient moisture to enable them to fall from the base of the cloud, they will fall as snow. In very cold air the individual snow crystals will retain the attractive six-sided shape that adorn ski brochures, but if the temperature of the air nearer the ground is above freezing the crystals will begin to melt. These sticky crystals amalgamate and fall as the big soft flakes typical of 'English' snow.

Hail is rain which has been frozen through being lifted by upcurrents to higher and colder levels. It will fall as soon as it is heavy enough to sink faster than the speed of the upcurrent. If, however, the upcurrent is sufficient to sustain it for some time while it is both falling through the cloud and growing it will become perhaps $\frac{3}{4}$ in. across. If it is now carried up again and the process repeated it will become really large. The best chance of this happening is when the

Rain, rain, rain

In half an hour enough rain to flood streets and houses fell on the Welsh coast town of Aberystwyth at tea time on a Saturday afternoon. If looked at superficially the synoptic chart for the day gives little indication of circumstances which could have been embarrassing to any small aircraft in the vicinity. The chart indicates that most of central and southern Britain was in the influence of a NE wind resulting from a Low over Europe. The deeper depression off Ireland would obviously be pushing in the cold front, but the pilot might be excused for thinking that its

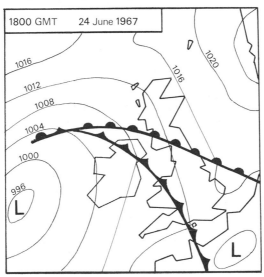

1800 GMT   24 June 1967

5.2

progress would be only normal and visible, thus giving him plenty of time to divert.

However, there were thunderstorms over Britain, showing that the air was unstable ahead of the cold front, as well as behind it, where the air was cool and very moist. At about 4 o'clock the easterly wind dropped, and after a short calm swung violently to the SW (the isobar direction behind the cold front), bringing in a mass of cloud accompanied by thunder and heavy rain. The cloud was obviously very thick because it became almost dark on virtually the longest day of the year; and it came right down and covered the hills. Like most unusual occurrences several factors were involved; the front had been moving uninterrupted over the sea with little to disturb it. On approaching the coast it was probably accelerated at least in the lower levels due to the vast masses of rising air over the land during the late afternoon; on arrival it would be immediately forced up over the hills and mountains causing a massive release of energy through condensation. If it had approached even a few hours earlier before the thermals had become really extensive and strong over the land, or later when convection had started to die down, the rainfall would

—continued

probably have been quite ordinary. However, a study of the charts for the previous day shows that the depression was in a strong and active enough state to make a considerable impact on the existing weather. This shows the value, where possible, of studying a sequence of charts, since flying near Aberystwyth during this afternoon would not only have been unpleasant but made hazardous by the high ground on the 'escape route' from the area.

cumulo-nimbus anvil (page 147) is streamed along by strong high-level winds. The hailstone is not only sustained in the cloud mass, but when it does fall it is likely to do so into the rising air at the leading edge of the storm instead of into the cold air downfall at the rear. It will now be carried up again and the process repeated. This ends when the hailstone has become so large and heavy that even upcurrents of 5000 fpm or more cannot sustain it. By this time it will have grown as large as a golf ball, even a tennis ball, and weigh over a pound. Whether it falls on aircraft, vineyards, or people, the damage will be expensive to repair. In some countries where the per acre value of crops is high, explosives are fired into storms in an attempt to shatter large hailstones, although there is no evidence that this activity is successful.

The sort of rain, hail or snow that falls can give us a great deal of information on what is going on in the dullest of leaden skies. Snow, for example, is most likely to come from thick layer clouds, such as occur in a warm front when the uplifting of air is only gentle. Hail forms only when there are strong upcurrents and convection clouds. Snow falls only in cold weather, but hail is possible at any time of the year, even in the hottest of summer weather. If the ground is warm it will rapidly melt, but in winter successive falls of hail will lie and look like snow over the land. On inspection they will be found to consist of tiny pellets of frozen rain, and not crystals. We can take it as a general rule that the arrival of hail or frozen rain means that the precipitation will last for a shorter time than when the fall is of genuine snow crystals.

If it is extremely cold the air may be unable to absorb moisture as water vapour, and instead it freezes directly into tiny particles that make the air glitter. This is appropriately called diamond dust, and is cometimes met with on high ski slopes when moist air rises out of a warm valley. It does not reduce visibility so much as add to the glare, and cause optical phenomena such as mock suns.

Rain, hail, and snow may evaporate before reaching the ground, or may be seen as a wispy veil hanging beneath the cloud, called virga. If the veil is of snow or ice it will look white even in the cloud's shadow, but greyish if composed of water droplets.

When flying in cloud in which there is a large quantity of super-cooled moisture, condensation will take place on the surface of the aircraft and form ice. Moisture will continue to freeze on to this ice, which builds up increasingly. The parts of the aircraft which are most susceptible to icing are the leading edge of the wings and tailplane, and the carburettor (page 198).

Recognising the clouds

When we say that clouds come in an infinite variety of shapes and sizes, this is the right adjective to use. No two clouds have been exactly alike since weather began. Nevertheless, they can be grouped for better identification, because clouds differ in appearance according to the behaviour of the air which created them. They also look different depending on whether they are made of ice crystals or are composed of water droplets. Sometimes it is easy to recognise a cloud – such as a summer cumulus – but more often there are masses of different clouds jumbled together at the same time. It is therefore difficult to sort out exactly what they are or why they have developed. To ease this problem the next few pages list and illustrate typical examples of representative clouds for reference purposes. Later in the book, and of course in the sky, more subtle variations of these examples will be found. In the book they will be looked at to explain why they have appeared at all and why they are the shape they are; but outside you will have to work this out for yourself.

Height and composition	Shape					
	Fibrous	Flat	Patterned	Heaped	High-Heaped	Wave
High 20,000–40,000 ft Ice crystal	Cirrus (Fig. 5.3) Contrails (Fig. 5.15)	Cirro-stratus (Fig. 5.4)	Cirro-cumulus (Fig. 5.5)		Cumulo-nimbus (Fig. 5.12) Mammatus (Fig. 5.13)	Al-cu-lenticularis (Fig. 5.14)
Medium 8000–20,000 ft Water droplet	Contrails (Fig. 5.15)	Alto-stratus (Fig. 5.6)	Alto-cumulus (Fig. 5.7) Al-cu castellanus (Fig. 5.8)			
Low 0–10,000 ft Water droplet		Stratus (Fig. 5.9)	Strato-cumulus (Fig. 5.10)	Cumulus (Fig. 5.11)		

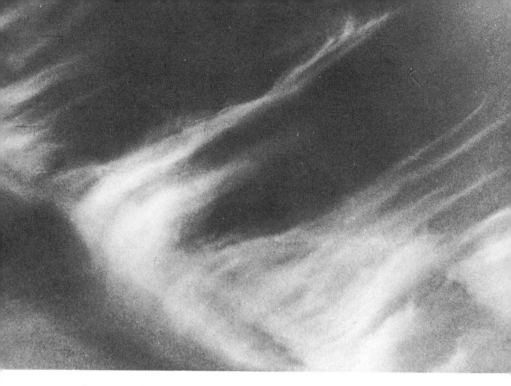

5.3

CIRRUS   (Ci) High, fibrous.

Detached delicate-looking clouds with filaments or a feathery pattern.

Usually white. Haloes with red on the inside of the circle are sometimes seen round the sun.

Height: 20,000–40,000 ft.

Composition: Ice crystals.

Cirrus cloud forms at heights where the air is so cold, at least −40°C,
that it is composed only of ice crystals and not of water droplets. This is
what gives it the feathery and delicate appearance. The gently ascending
air of a warm front reaches cirrus levels and the development of this
cloud may be the first sign of an approaching depression. Cirrus also
appears in anticyclonic weather but is more patchy. Cirrus in long
streamers is evidence of strong winds, sometimes of a jet stream. If the
cirrus fingers can be seen from the ground to be visibly shifting across
the sky, the wind is probably blowing at 100 knots or more at that level.

(Chapters 5, 7, and 12.)

5·4

CIRRO-STRATUS   (Ci-st) High, flat.

A thin, fairly uniform layer which does not blur the outlines of the sun or moon, but causes haloes to appear around them.

Translucent white.

Height: 20,000–40,000 ft.

Composition: Ice crystals.

When the sky is covered with a thin filmy sheet of ci-st which is gradually thickening, a slow or steady deterioration in the weather may be expected, because the cloud is developing as the result of expansion and uplifting of air over a very big area. It is typical of the sky associated with an approaching warm front, following the earlier cirrus. The front is still far away and the cloud is still high enough to be formed only of ice crystals.

(Chapters 5, 7, and 12.)

5·5

CIRRO–CUMULUS   (Ci-cu) High, patterned.

A thin white cloud which shows a pattern in the form of lines or ripples. Similar to Cirrus but instead of being fibrous each line or patch is composed of a large number of very small blobs of cloud.

White.

Height: 20,000–40,000 ft.

Composition: Ice crystals.

Ci-cu are not common, perhaps because they are very short-lived. They may occur in a patch or sheet of cirro-stratus which is decaying. They are of little significance, but beautiful.

5.6

ALTO-STRATUS   (Al-st) Medium levels, flat.

A more or less uniform layer of cloud which is obviously lower than ci-st.
The sun and moon show through vaguely.

Greyish.

Height: 8000–20,000 ft.

Composition: Water droplets.

The damp-looking sheet of alto-stratus usually means that rain of the
drizzly variety has already started to fall from the cloud, but is evaporating
before reaching the ground. When it appears ahead of an approaching
warm front there is every likelihood that the weather will worsen before
it starts to improve. If there is no indication of the location of the sun
through the cloud it can be reckoned to be several thousand feet thick.
The weather will usually deteriorate faster in hilly country.

(Chapters 5, 7, and 12.)

5·7

ALTO–CUMULUS   (Al-cu) Medium levels, patterned.

Cloud in a fairly thin layer with a pronounced small scale pattern.

White or greyish.

Height: 8000–20,000 ft.

Composition: Water droplets.

Often called a mackerel sky. Alto-cumulus clouds are often a sign of clearing skies, so their appearance does not usually indicate approaching bad weather. They are mostly seen late in the day when a sheet of cloud is breaking up to give a fine night. They are attractive to look at and especially so when the low sun lights the underneath of the cloudlets.

5.8

ALTO-CUMULUS-CASTELLANUS   (Al-cu-cast) Medium levels, separated.

Small clouds without uniform distribution. Noticeable for their height in relation to their size, sometimes giving the appearance of standing on their tails.

White.

Height: 8000–15,000 ft.

Composition: Water droplets.

Precursor of thunderstorms. These little puffy clouds do not usually last long but indicate instability in the medium levels. They are often seen earlier in the morning than ordinary cumulus would be expected and disappear sometime before the thundery conditions develop or move in. They sometimes have a pearly appearance.

(Chapter 9.)

5.9

STRATUS　　　　(St) Low, flat.

A uniform layer of cloud. Where the cloud lies on the higher ground it is termed hill fog.

　　Grey.

　　Height: 500–5000 ft.

　　Composition: Water droplets.

Stratus may develop due to air being raised and cooled as it flows in from the sea, or it may be caused by an approaching warm front. If associated with a front the cloud will be soft-edged and weepy without a distinct base, and will gradually lower until it is on the ground. Stratus coming in from the sea will tend to lift and break as it travels further inland. When stratus is decaying or the base is lifting the bottom of the cloud will be better defined with improving visibility below.

(Chapters 5, 7, and 12.)

5.10

STRATO-CUMULUS   (St-cu) Low, patterned.

Low cloud in a fairly thin layer with a pronounced pattern, sometimes in lines.

Grey.

Height: 500–5000 ft.

Composition: Water droplets.

A common cloud in cool damp climates. Strato-cumulus may start as fog or a low sheet of stratus which lifts during the morning as soon as the sun starts to stir up the surface layers of air. Depending on the temperature and humidity it may break into the typical globules shown here or change into genuine cumulus, disperse entirely or persist all day. It frequently clears at night.

(Chapters 4 and 12.)

5.11

CUMULUS   (Cu) Heaped, separated.

Single clouds with some vertical thickness. The upper surface is often dome-shaped, and the base flat, horizontal and shadowed.

White on top, darker underneath.

Height: Base 1000–15,000 ft (Britain seldom above 7000 ft).
Tops 1200–25,000 ft (Britain seldom above 12,000 ft).

Composition: Water droplets.

The cloud of the glider pilot and the summer scene. With a thermal underneath, or when not too large to keep the sun away, they head the popularity poll. But cumulus clouds can provide a great deal of information about the quality and behaviour of the air. In dry settled weather they will be small and flat, but larger and with lower bases when the air is moist.

(Chapters 7, 8.)

5.12

CUMULO-NIMBUS   (Cu-nb) Heaped, large.

Big clouds with great vertical development, often with tower-like summits. The top of the cloud has a fibrous texture and grows in anvil shape.

White on top, often very dark underneath.

Height: Base 3000–15,000 ft (Britain seldom above 7000 ft).
Tops 10,000–40,000 ft (Britain seldom above 30,000 ft).

Composition: Water droplets (lower cumulus), ice crystals (upper anvil).

Cumulo-nimbus grow from cumulus. If they become big enough so that their tops are in the low temperatures of the upper air, the cloud at these levels will be composed of ice crystals; this is why they have the two-part appearance.

Cu-nb have all the dramatic qualities, like a meteorological prima-donna; grand to look at, they have every mood in the book and un-believable power. Rain, hail, snow, thunder and lightning, darkness, strong and gusty winds, heat and cold are all in their repertoire.

(Chapter 9.)

5.13

MAMMATUS   (Mam) Bulbous.

Globules or festoons which droop from the base of big cumulus and cumulo-nimbus clouds.

Grey or dark grey.

Height: 1000–15,000 ft.

Composition: Water droplets.

There is an ominous look of impending doom in these clouds as they bulge downward from the black sky overhead; but they are at the tail of storms and usually pass on their way, or decay, fairly quickly. Heavy rain or hail may fall from the same storm cloud. Because mammatus occurs only locally under part of a storm it can easily be flown around.

(Chapter 9.)

5.14

ALTO-CUMULUS-LENTICULARIS   (Al-cu-lent) Smooth edged.

Clouds of ovoid or lens shape often with clean cut edges. They occur at all levels and generally remain stationary in the sky.

> White or greyish.

> Height: Up to 50,000 ft but unlikely to be less than several hundred feet above the ground.

> Composition: Usually water droplets, but crystals high up.

Signposts to great waves in the sky. Appearing usually to the lee of mountains and big ranges of hills they indicate the presence of torrents of air which can drag down an aeroplane or sweep it up thousands of feet. The lenticular clouds form near the crest of each wave growing continuously on the upflowing side, and evaporating continuously in the downgoing air. This is why they remain stationary over the ground. An attractive cloud which is more friendly when understood.

(Chapter 11.)

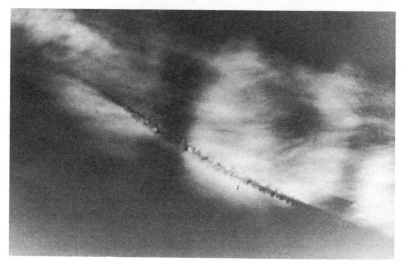

5.15

CONTRAILS   Fibrous, in lines.

Lines of condensation resulting from the passage of an aircraft. Initially the trails are narrow but sometimes spread out into broader bands.

White.

Height: Usually above 10,000 ft, often above 20,000 ft.

Composition: Water droplets, but ice crystals high up.

If the air is dirty and damp, contrails will persist and sometimes develop into genuine cirrus cloud, occasionally spreading to cover large areas of sky. In drier air their appearance may be only intermittent, if they appear at all. Occasionally they cause evaporation of the cloud.

6 Clouds in quantity

Air mass clouds

Depressions and anticyclones give us clouds and weather which are recognisable in both sequence and appearance. High-pressure conditions are usually pleasant for flying, but low-pressure systems – depressions – are capable of providing weather ranging from merely dull to disastrous.

When airborne, it is of less importance that the bad weather ahead has or has not been forecast, than that we can recognise it for what it is. If a great black wall of cloud stretches from horizon to horizon across our track there is only one proper action, and that is to hurry back the way we came; the decision is easy to make because the danger is obvious. It is when the weather deteriorates gradually and all the time our destination is getting closer, that it is more difficult to take the correct remedial action. Unless we are able to recognise what is happening, anticipate the rate of deterioration, and work out a safe alternative, we should regard all depressions on track as an insidious menace.

6.1 Approach of a warm front

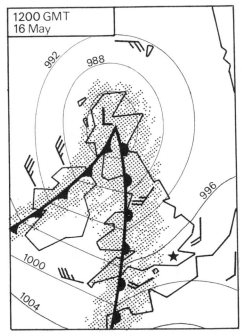

1200 GMT
16 May

992
988
996
1000
1004

This map should be studied with the photographs overleaf.

A

B

C

The first sign of an approaching warm front is cirrus creeping up from the horizon (**B**). If the depression is moving in quickly, with strong high level winds, the first cirrus may be seen as fingers stretching out ahead of the cirro-stratus (**C**). Figure **B** is what the front on the map would look like to someone at * if it were a small depression, but the influence of a larger and deeper one would have already brought complete cover as in (**D**). Photographs. 1130, 1230, and 1600 hrs. May.

D

E

While the frontal cirrus is still low on the horizon the day may seem fine and settled. We should not be misled, but watch the rate of advance of the cirrus across the sky, and observe whether the approach of the front is towards us or whether it appears to be sliding along the horizon. As soon as the cirrus moves across the sun deterioration in the local weather will be more rapid, and will be accelerated as the declining sun sinks down behind the thickening cloud.

F Hill fog
This photograph was taken shortly after a warm front had cleared the area. The rain has ceased and cloud base has lifted. Half an hour earlier cloud was in the valleys, and everything was obscured by drizzle. Height of these hills 2000 ft. Wind west, strong and gusty. July.

F

Although we may find ourselves flying in any part of a depression, and at any stage of its life – it may even be born around us – it is more logical to describe the cloud sequence of a typical depression from the appearance of the first cirrus, through to the polar air weather behind the cold front (Figs. 3.4 and 3.5). So to get a picture of what to expect let us assume that we have a quiet, fine day with just a few untidy cumulus scattered about; over the last few days we think we remember having seen some jet stream cirrus, and that cloud and rain is forecast. Ahead of the still distant warm front we know that air has been gently ascending over an area of perhaps 30,000 square miles, rising over the cooler air ahead of it – the air we are in. The leading edge of this relatively warmer air is at about 30,000 ft and at a temperature of perhaps −50°C. From here right back to the front itself there will be cloud, all over this vast interface, which is progressively lowering and thickening. The highest clouds will be ice-crystal cirrus, and the lowest water-droplet stratus. From about 200 miles away it may be possible to see the cirrus near the horizon, either as wisps or as a sheet which gives the sky a wan or milky appearance, becoming yellower than usual towards sunset. If a barometer is available it will show slowly falling pressure; if an aircraft altimeter is used as a barometer, it will indicate increasing height at the same point on the ground. If we are using it as an altimeter, it will show height above the ground which we do not have. Although it will not rain for some hours yet, the time available before it does so will depend on the rate at which the cloud is moving in.

If there are cirrus fingers stretching out ahead of the front, this indicates that winds are strong high up; if they can easily be seen from the ground to be moving, the wind at around 30,000 ft is blowing at something like 100 knots, and the front is probably coming in fast. If, however, the leading edge of the cloud is more like that of a veil and its movement can only be discerned after lengthy observation, the upper winds are not so very strong. If the edge of the cirrus has a more solid appearance, or is crumbling into little globules of cloud, the front is decaying and the depression may have stopped moving. Even if it is still creeping in very slowly, it may be decaying sufficiently fast for the rain to cease before it gets to where we are. The pressure should be checked from time to time to see whether the pre-frontal fall is slowing or has stopped.

If the warm front is steadily moving in, the advance feathery cirrus will soon, but almost imperceptibly, give way to a more uniform cloud cover. The sheet will still be translucent, with the sun remaining fairly bright. If no warning has been received that a front is on the way, the signs, so far, could be easily missed, since the weather near the ground will not yet have undergone any noticeable change. When the sun is surrounded by a halo which has red on the inside of the circle, all the cloud in the sheet is high enough to be composed of ice crystals.

If we can see the cloud working its way steadily up the sky, it is obviously moving in our general direction, but if it appears to be sliding along and staying near the northern horizon, the centre of the Low will pass to that side; consequently our locality will not receive the full influence of the depression. An indication of the direction of movement can be obtained from the local wind, both on the ground and at high levels. There is a rule – Buys Ballot's Law* – which says that when we are facing the wind, in the Northern Hemisphere, pressure will be lower away to our right.

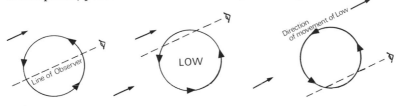

6.2 Wind direction experienced by an observer will vary with the path taken by a depression. Isolated arrows show direction of movement of the Low; those on circles the wind direction round the depression.

In the early stages of the approach of a Low, there is likely to be some backing of the prevailing wind, with a subsequent veer at the passage of both warm and cold fronts.

If when the first signs of cirrus appear in the west, or SW, sky there has been no recent backing, or alternatively if there is any veer, the centre of the depression is passing more to the north. If the wind backs from, say SE through East to NE, the centre of the Low is passing to the south. Normally the wind veers with height, but changes in the flow direction can be best seen by comparing the movement of the high cirrus with that of any lower cloud,

* His law is actually worded a bit differently, since he insisted on standing with his back to the wind, and having the low pressure on the opposite side; this is not very sensible because we would be unable to see the weather that is approaching, which is the main point of the exercise.

such as small cumulus, or the wind on the surface. The upper wind gives some indication of the possible path of the whole depression, but should not be given too much importance if the cloud is a very long way ahead of the Low centre.

However, we will assume here that the depression is moving in toward us. As the warm front approaches, the interface between the warm and cool air will continue to lower, and the cloud in this thickening sheet or sheets will descend to the levels where it is made of water droplets. By now the sun will look pale and indistinct. Gentle rain may already be falling from the cloud, but evaporating before reaching the ground. Without the full warmth of the sun, the day will feel cooler, and any cumulus will flatten and soon disappear. The dull cover of layer cloud will steadily thicken and become lower until the front arrives. This moment will be difficult to discern because of the drizzle which will have been falling for some time. Rain from layer or sheet cloud is composed of very small droplets, which cling to the windscreen, and are difficult to see through. Hilly country will have more extensive low cloud, or hill fog as a result of the damp air being forced up over it; condensation will take place, often in seemingly random, or irregular, patches. The unevenness of this *Orographic* cloud, as it is called, is due, not only to the air being pushed up over the hill, but because the turbulence creates little pockets of lower pressure in which rags of cloud form. Orographic cloud is dangerous because it develops quickly, randomly, and obscures ground already indistinct in the mist and drizzle. Getting lost in this sort of weather is something to avoid.

The warm front may take two or three hours to go through, with cloud base either on the ground, or obscured by drizzle.

The pilot said . . .

. . . that 10 minutes after take off he encountered low ceiling and visibility and turned back towards the airfield from which he had just departed. During the turn visibility deteriorated to near zero. As a result he deviated into mountainous terrain uncertain of his precise position. Suddenly he realised he was near tree top level with sharply rising ground ahead. The aircraft hit trees with its left wing, spun around, and crashed into deep snow.

The warm sector

After the front has passed the prevailing air is 'warm' (Fig. 6.3). There may be a slight veer in the wind, a small rise in temperature,

6.3 Warm sector weather

A

B

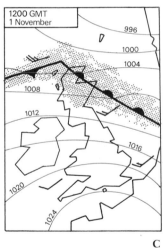

C

Weather in the warm sector of a depression may vary considerably. Near the low-pressure centre it can be cloudy and windy, away from the centre there may be blue skies with cumulus, and little or no high cloud.

In general warm sector air should be expected to be quiet and cloudy, usually with a fairly low cloud base, since the air is frequently moist. Visibility is usually reasonable, and cloud thins and often breaks during the evening and night.

A 30 April. 1645 hrs. Wind, 12 knots. B August, 1200 hrs. Wind W, light.

D

E

D The difference between alto-cumulus and strato-cumulus is sometimes only one of height. These clouds are high strato-cumulus; the sheet is breaking up and the sky cleared shortly afterwards. Wind WSW, light.

E A sheet of stratus clearing in the early evening, with decaying alto-cumulus above. The warmth of the sun had caused enough stirring up of the surface layers to create the stratus, now it is collapsing in the cooling air. In winter the clearing sky would be marked by an appreciable temperature drop due to radiation. July. 1900 hrs. Wind W, light.

and the pressure will cease falling. Whether or not it stops raining will depend on the humidity of the warm sector air, and the temperature of the surface. If the air has come from over the sea and is very moist and salty, and the surface of the land over which it rises and travels is cool – and it may be since it will be soaking wet and trying to evaporate itself dry again – there will be little incentive for the rain to stop. This sort of situation is by no means unusual on Atlantic facing coasts.

Often warm sector weather is quiet, muggy, and with a good deal of cloud, usually stratus or strato-cumulus. In the north European winter it gives a typical grey day. In summer, however, if the warm sector is large and the air fairly dry, it can be fine and warm, with cumulus clouds and good visibility. If the warm sector stretches for a hundred miles or more, pressure may rise considerably and the weather settle; if there is a big anticyclone adjacent it may bulge or grow into this area. If this happens, whatever is left of the old depression will be further weakened. A large warm sector in a slow moving depression may last several days, so if the centre of the Low is also moving away, or if it is declining, the cold front may never arrive. However, in the more usual situation the warm sector will be terminated by the arrival of the cold front.

The cold front

A cold front (Fig. 6.5) always has a feeling of violence about it, even when it has become occluded and mixed up with the gentler warm front drizzle. At its most extreme it will travel across country as a gigantic line of squalls, but in geographical situations which break up the smooth flow of weather, as Britain does, the cold front is rarely as powerful or dangerous, although its weather may be frequently rough and unpleasant. Over land masses such as the United States, with large, relatively uninterrupted north-to-south distances and a wide range of temperature, cold front squall lines of remarkable intensity develop. Occasionally cold Canadian air sweeps 1000 miles south into a Texan summer, dropping the local temperature by 10°C or more. As it travels southwards the cold air is meeting up with increasingly warmer air; up goes this warm air along the front line, allowing the dense cold air underneath to charge along at speeds sometimes as high as 50 knots (Fig. 6.4).

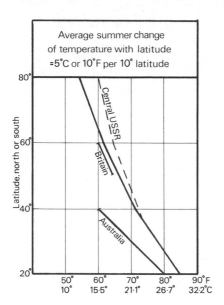

6.4 The average change of temperature with latitude is enough considerably to modify weather which is moving or likely to move over large N–S distances. A cold front, for example, sweeping down from Canada towards Central America may change its latitude by 20 deg., and temperature anywhere between 5 and 10°C. This extra warmth, or cold, is available to increase the ferocity of a line squall, or alternatively turn harmless rain into a blizzard.

The cloud that develops along a cold front line will include quantities of cumulus and often cumulus-nimbus, so the rain which falls will be more of the shower type. It is likely to be sudden or heavy, and with hail and thunder. The droplets of rain will be considerably larger than those which descend so dismally from a warm front. The belt of cloud will be far narrower than that of its predecessor, with, therefore, much less warning of its approach. The height of typical cold front and line squall clouds is roughly between 8000 and 15,000 ft, and an indication of the thickness of cloud is the extent to which it becomes unusually dark for the time of day.

The cold front usually passes quite quickly, and following it the air will be cooler, with a cleaner feel and some veering and freshening of the wind; a barometer will show an abrupt pressure rise, and an altimeter set at airfield height will under-read; that is, it will indicate that the aircraft has reached airfield height while it is still some 50 ft up.

6.5 Cold front

The approach of a cold front squall line. Such storms may travel faster than the mean wind speed. Closer to the cloud the air is likely to be very turbulent with gusts blowing outwards from the region of the rain. More often cold fronts are mixed up with, and obscured by, other cloud, particularly in maritime and hilly countries.

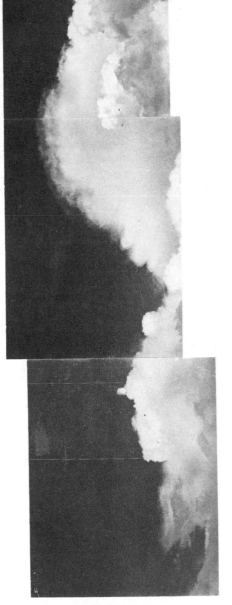

A July. 1430 hrs. Wind NW.

6.6 Cold air

The weather map (D) shows a cold front lying across Britain, which would bring weather similar to that in the photographs opposite.

A and B show how the sky looks in the cold air behind the front. (C) shows how the weather becomes quieter and more settled further behind the front.

On the chart the change in wind direction and the increase in wind speed with the passage of the cold front can be seen clearly.

A

May, 1625 and 1630 hrs. Wind WSW, 15 knots.

B

C

July, 1600 hrs. Wind WSW, 10 knots.

6.6 *contd.*

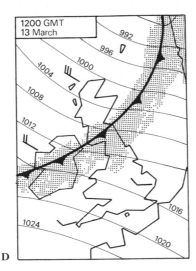

D

Occlusions

Depressions do not often arrive over west Europe in a textbook fashion. They are more likely to be partly occluded, with the faster moving cold front air having caught up with, or even having moved through, the slower warm front. It may be difficult to discover if this has happened since formless cloud and drizzle usually obscures all. However, if both fronts have come together invisibly above, the cloud along the occluded front will have become thicker where the convection cloud merges with the warm front mass; the sky darkens, the light worsens, and the character of the precipitation changes. The drizzle gives way to the heavier rain, which may also feel colder, or fall as sleet or snow. At the rear of the cold front cloud-line the rain often becomes very heavy for a short time, so that any increase in the amount of the rain and the size of the drops heralds the end of this particular cloud mass. Through the rain it may be possible to discern the first pale glimmer of clearance along the upwind horizon. The clearance, however, may not last very long.

An occlusion may be warm or cold, depending on the relative temperature of the air mass ahead of the warm front and of that behind the cold front. This latter air will obviously be colder than the warm sector air under which it is sliding, but it could be warmer than the air ahead of the warm front, in which case on reaching it the air will rise. If, on the other hand, the cold front air finds air

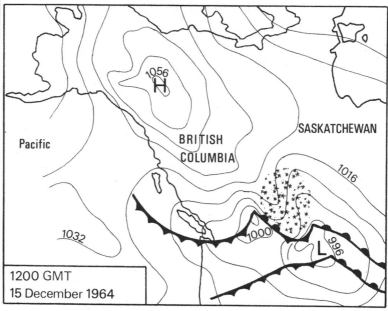

Blizzard in Canada

SASKATCHEWAN

Pacific

BRITISH
COLUMBIA

1056
H

1016

1032

1000

L
996

1200 GMT
15 December 1964

6.7

To count as a blizzard, the Canadian met service requires a storm to:

(a) last 6 hours or more
(b) blow at 25 mph or more
(c) have a temperature of 22°F or less, and
(d) have a visibility of half a mile or less.

The storm shown on these synoptic charts fulfilled in good measure all the requirements. Fortunately a warning was put out in good time so no one died through being caught out, although some ranchers lost up to half their cattle and over half the pheasant population of Alberta were killed off. In Saskatchewan visibility went down to 200 yds in blowing snow with a wind of 55 mph. There were gusts up to 70 mph, the temperature sank to 4°F, and these conditions lasted for a day and a half!

For some days West Canada had lived with sub-zero temperatures, and a depression moving down from the NW showed every sign of bringing trouble. As it approached a second Low sidled in from over the sea, merging with the Northerly one, and deepened some 16 mb into a violent and complex disturbance. At 12 GMT on 15 December the blizzard covered the shaded area of over 100,000 square miles.

6.8 Alto-cumulus

Alto-cumulus results from gentle
convective instability within a
shallow layer in the higher levels.
If there is any wind sheer the
cloud develops in rolls or bars
(A). Sometimes al-cu thickens up
into a sheet of alto-stratus, but
more often, particularly late in
the day, the reverse happens;
the sheet of cloud is thinning and
A decaying.
    Al-cu has no special
significance as a precursor of
anything dramatic happening to
the weather. It tends to appear
more when the weather is
beginning to quieten down –
even if only temporarily. To the
balloonist it may offer some
prospect of suitable conditions,
but for the glider pilot it is simply
an undesirable obstruction
between the sun and the ground.

—continued

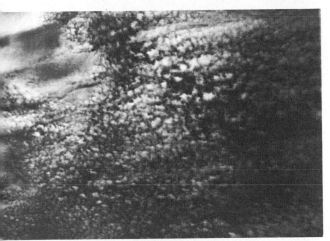

B

C

D

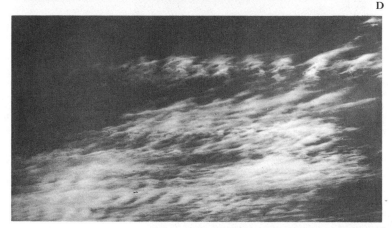

E

6.8 (D and E) Al-cu developing in conditions of considerable wind shear. The wind was strong. The clouds were not only moving rapidly across the sky, but constantly changing in shape, and in the amount of cloud present. Al-cu of this appearance should be regarded with suspicion as cloud may increase really rapidly to complete cover without warning.

F Al-cu forming in rolls due to wave undulations at wind shear level. The bars lay across wind and were, of course, moving at wind speed. In neither case did they last more than a few minutes. They should be regarded as warning of strong winds at high level and the probable approach of unsettled weather. The addition of the cirrus (G) is evidence of winds of up to 100 mph at around 30,000 ft.

F

G

Winter thunderstorms in central Europe

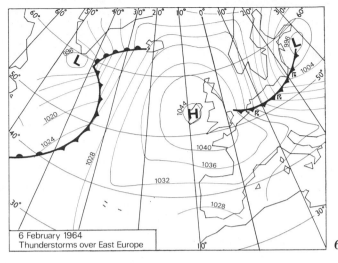

6 February 1964
Thunderstorms over East Europe

6.9

Although thunderstorms are usually thought of as summer phenomena, there are sufficient thunderstorms in winter in Europe to be taken into account when planning a flight in certain synoptic situations.

They are likely to occur when cold sea air flows from the NW – down the North Sea and from over Scandinavia – and into central Europe. The illustration shows a typical situation with a cold front over the Baltic between a Low over Russia and a High over Ireland. This High does not help matters as it is squeezing the northerly flow and so increasing the strength of the wind. The cold front which is over 800 miles long, plus any following squalls, sweeps southwards at strong wind or gale force speed. The storm does not last longer locally than 10–15 minutes but in passing produces thunder, lightning and heavy snow. There is likely to be a temperature fall behind the front of some 5°C, and a veer in the wind direction. The air after the front and between any following squalls is clean and clear.

In a light aeroplane we should regard such a frontal barrier as impenetrable, and fly out ahead of the storm, downwind, and land at some airfield where the aircraft can be put under cover until the front has gone through. However, the pilot of a glider who is impervious to cold might well find that flying along the leading edge of such a front provided an exhilarating ride.

Although summer thunderstorms may extend up to 30,000 ft or more, it is unlikely that winter thunderstorms in the cool weather areas of the world go much above 13,000 ft.

—*Aero Revue*. April 1969, p. 205

warmer than itself on penetrating the warm front, it will continue sliding along underneath (Fig. 6.6).

Although we talk of cold and warm air in these frontal situations, the difference in temperature may be as little as 1°C. When an old occlusion from which rain has ceased to fall passes overhead, the change in air temperature may not even be noticeable. Sometimes, of course, the change will be much greater, as in the fast-moving cold fronts mentioned earlier hurtling in the direction of the sub-tropics over a big land mass. Sometimes a front, under the influence of the environmental pressure pattern, is stopped or even caused to retreat. If this happens the weather will change according to the way the air is forced to behave. If, for example, a weak cold front starts to reverse, the warm sector air will float back up over the cool, and give warm front type weather going the other way. This situations occurs not infrequently, and can give us succeeding days of cloud and rain on a swinging wind direction.

As a depression ages the occluded front will cease to give rain, and the cloud will lift and become thinner, drifting for a while in the medium levels before evaporating away altogether. The passing of an aged occlusion may be noticed only by a small shift in the wind, a slightly different feel or smell to the air, and a few hours of cloud which does little more than reduce the warmth from the sun.

Clean and cold air

The air mass behind an active cold front is usually clean, cooling at the dry adiabatic lapse rate of 3°C per 1000 ft, and probably moving over wet ground. The effect of strong sunshine is to heat both the surface and the air in contact with it, creating an unstable situation. Quantities of warmed air expand and rise, producing cumulus, often large enough to give heavy showers. These conditions can be very frustrating, with spells of brilliant blue sky alternating with great chunks of dark cloud spewing out rain or hail. Such rapid successions of storms, or mini fronts, may well continue until the energy of the day subsides with the sun – and they will usually become less potent with a dropping wind.

It is not always appreciated how much difference the arrival of clean air makes to the amount of heating received by the surface. The air of maritime and largely urbanised lands is dirty, containing nuclei, salts, and dust. The haze produced by these particles reflects

Unforecast snowstorm

Up the Yellowstone River canyon, through Yellowstone Park, and across the Continental Divide to Jackson. There were to be scattered cumulus along my route and 10 knots of westerly wind, with an occasional light rain shower, but I would call for the latest weather just before entering the narrow Yellowstone valley.

Sure enough, I ran into a rain shower. As predicted, it was no problem and I could see through it without difficulty. . . . A little while later I entered another precipitation area and was surprised to find that it was light snow. I encountered several such flurries but they were only troublesome enough to keep me from completely enjoying the mountain scenery. . . . I took several pictures of the Yellowstone Falls before continuing southwards. I was flying at 8500 ft when I reached the lake, some 800 ft above the surface. Suddenly snow started falling, this time a good deal harder than before. I took a look behind me; snow was falling there too. The only thing I could make out clearly was the terrain directly below me; navigation was restricted to following the lake shore.

When I reached West Thumb on the western edge of the lake it was snowing as hard as ever. I had three choices. I could see Route 89 so could stay low and follow the road through the pass to Jackson, or I could continue flying around the lake and wait for the snow to stop. Thirdly, I could climb to 12,000 ft or so and get out on instruments. Landing below was an impossibility.

I decided to follow the road at 9,000 ft in the hope that I would fly out of the storm; if I lost sight of the road for more than a few seconds I would turn to 210° and begin climbing. I don't suppose I got more than two or three miles before Route 89 vanished in a swirl of snow. The magic altitude was 10,308 ft which would put me above Mt Sheridan. As I passed through 10,000 ft my blood suddenly ran cold. There was ice on the windshield and the elevator. I felt a strong urge to panic. The situation was that I was on instruments below the level of nearby terrain and unsure of my position, there were no radio navigation facilities available, no instrument flight plan or clearance, and no communication with anybody. I was near the maximum altitude at which I could hope to operate for very long without oxygen; there was icing at an unknown rate and I had no previous experience in icing conditions. The only thing I could do was to hang on. Evidently the icing layer was not too thick and I made it to 12,000 ft. I now occasionally broke cloud between layers and most of the ice began to melt off. . . .

Approaching Dunoir the weather cleared a bit, and I started to let down through a large hole. As I began to do so the ASI wound up and hit the peg at 180 mph. Then I happened to think that the airplane just didn't sound like it was going 180. I peered out of the window and saw

—continued

a considerable amount of warmth back into space, but when clean
air moves in the heat available to the surface increases dramatically.
The day does not necessarily feel hotter because clean air is often
accompanied by a wind, and the air in the shadow of the big clouds
can be cold. In spring particularly, such days can be wild. The
effective heat of the sun is increasing daily, but the ground and air
are still cold from winter. Quantities of surface moisture are sopped
up into the sky, turned into tall clouds, and quickly rained back again.

During its passage across any broken or variable surface, such as
is provided by the British Isles, the approaching air mass, its
circulation, and any fronts will be altered in character and behaviour.
The importance of this when flying is that the weather locally may
become markedly different to that in the general forecast. The cold
air of winter, particularly, may produce some completely unflyable
local weather. Windward facing hills can become solid with orographic cloud and snow, and moisture will be quickly wrung out of
the air by the high ground. If warm frontal air moves in over ground
and air which is very cold, the drizzle will freeze on the way down
and fall either as snow, or more often, as tiny pellets of frozen rain.
As the warm air moves into the lower levels, the precipitation often
changes back to rain; so if it is necessary to fly in this sort of weather,
we may be better off nearer the front line in the rain, than further
ahead of it in the snow; either way engine icing is a risk, but visibility is better in rain than snow. When the ground is extremely cold
our troubles may not be over, even on landing, since any rain will
almost certainly have frozen on impact with runway and road surfaces. Ice formed by this contact freezing is usually clear, and therefore difficult or impossible to see.

In cool, damp countries with a considerable north-to-south length,
such as Britain or Norway, there may be a considerable temperature
difference between one end and the other, so a change on the way
from rain to sleet or snow, or vice versa, should be anticipated. The
change may take place imperceptibly, but in such weather it is easy
to become lost, and run out of the short winter daylight while still
trying to locate an airfield.

7 Settled weather

When pressure rises, the weather becomes more settled (Fig. 7.1). If pressure rises really high, 1032–1040 mb, the weather is likely to stay settled for several days – perhaps even weeks, hot in summer and cold in winter. As we have seen, an area of high pressure – an anticyclone or a ridge – is a region in which the air is very gently subsiding. This causes compression and therefore warming of the air, and so there is little incentive for cloud to develop. In winter there is not likely to be much more than a layer of dull and overcast stratus, and in summer perhaps no cloud at all. If, however, there is hot sun and strong ground heating small local convection cumulus may appear. Away from the High centre the wind will steadily increase, and around the periphery it may be surprisingly strong, particularly in the direction of an adjacent Low.

The air does not subside uniformly in a High; it has to contend with convection from the surface carrying warmth upwards, and it may include patches or layers of air from which cloud has already evaporated. Subsidence and the effect of convection into the lower layers will create a level or layer in which the temperature slows or ceases to fall with increasing height; or it may even become warmer. If it simply ceases to fall it is termed an *isothermal* layer. If there is an actual rise of temperature with height it is called an *inversion*. It will be obvious from this that the dry adiabatic lapse rate of 3°C per 1000 ft will not be maintained in these circumstances, except possibly in a very shallow layer near the ground.

Whenever the temperature ceases to fall with height convection will be discouraged or inhibited; an inversion of temperature will in fact act like a lid trapping dust and smoke beneath it (Fig. 7.3). The higher the pressure the lower the inversion lid is likely to be. The sky will invariably remain clear of cloud, but the visibility, even in summer, may become so bad that visual flight is impracticable. Conditions will be worst at the centre of the High where subsidence is strongest, improving with increasing distance from the centre, except downwind of cities and other smoke sources. In this respect one disadvantage of an anticyclone is its persistence; the longer it stays, particularly in urban areas, the more squalor that

7.1 Anticyclonic weather

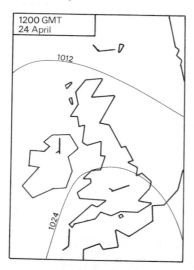

A 'All areas will have sunny periods and, apart from a few isolated showers in N. Scotland at first, and some coastal fog patches at times, will have a dry warm day'.

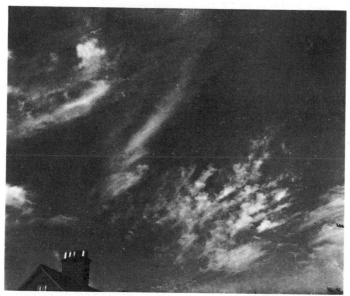

B Cirrus typical of anticyclonic weather and quite different from the thin sheet or streamers which develop ahead of a warm front. This cirrus is random both in pattern and direction.
June. 1400 hrs. Wind E, 15 knots.

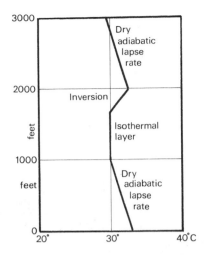

7.2 How an isothermal layer or an inversion modifies the rate of temperature fall with height.

becomes trapped and suspended in the surface levels of the air.

Temperature inversions can be seen clearly when flying. Below them the air is hazy, but above visibility improves immediately, with the top of the dust layer seen as a purple line around the horizon, darkest opposite the sun. We can usually see features on the ground quite well when looking vertically downwards, but much less well obliquely, especially into the sun. The haze from a big city may lie across the countryside for 50–100 miles like a huge grey stream, so we should avoid tracks directly downwind of such sources. Even a couple of miles away to one side may be enough to find vastly improved visibility. When flight planning, we should take into account the direction in which the anticyclone centre is likely to drift so that we will have an idea of possible changes in the lie of the haze streams.

In summer, if small cumulus start to reappear after a few cloudless days the inversion locally is becoming weaker. This may be because the anticyclone is declining, or that it is drifting away – so that we are becoming further from the highly stable centre. Because the air is fairly dry anticyclonic cumulus are usually small, flat, well isolated from each other, and short lived (pages 137 and 143). The appearance of cumulus is usually a sign that there will be some improvement in visibility.

Anticyclonic weather differs much more between summer and

7.3 Anticyclonic weather

A Above the inversion; dust is trapped in the surface levels and the top of the haze layer can be clearly seen. On this day visibility was quite good, but if an inversion persists for some days the dust accumulation may reduce visibility to a mile or less.

B and **C** Photographs taken from the same place show the 'burning off' of radiation fog with increasing warmth from the early sun. Complete clearance came within an hour. The weather was anticyclonic, and the day quiet, fine, and very warm in the afternoon.
September 0800 and 0850 hrs.

A

B

C

7.4 Alto-cumulus

A A sheet of al-cu which is deteriorating. The cloud within the triangular shaped area was slowly rotating, and disappeared more quickly than the remainder.

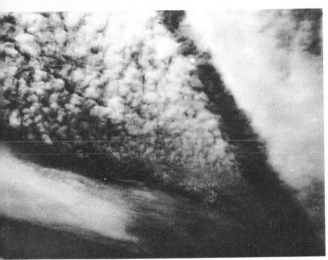

B These bars of cloud were more nearly parallel than appears. There was obviously some wave motion associated with wind shear at cloud level. The centre bar is decaying rapidly.

winter than that of depressions. In summer the stable air, small amounts of cloud, and light or steady winds allow the surface and the air in contact with it to become very warm. This is ideal holiday weather, and provides the hottest days to be found in Britain, particularly if the High is large with the centre over Germany, and pulling up air from the Mediterranean and North Africa. In winter

7.5 Cirrus

Patches of cirrus sometimes appear in the clear skies of anticyclonic weather. They may become sufficiently dense to detract from the warmth of the sun. Anticyclonic cirrus does not have any 'grain' to its pattern, and does not lie in lines. July 1945. Wind W, light.

anticyclonic weather may well be the coldest. For a start the air making up the system has probably been over NE Europe or even Siberia for some time. It then picks up a raw dampness as it flows across the Baltic or the North Sea, so that it is bitter when it streams across Britain and NE Europe. It quite often brings in a layer of low stratus cloud which shuts off the sun, so that such little warmth as there is will largely be reflected back into space from the top of the cloud layer. Should dust particles trapped in the lower layers be saturated or nearly so, the haze will turn into fog, ensuring that the amount of warmth now able to reach the surface will be negligible. The temperature on the ground even during the day may remain below freezing, and the fog and frost will persist until the anticyclone moves or decays.

An anticyclone decays when it is no longer fed by air subsiding into it from high levels, so that the loss from the divergent winds around the perimeter of the High causes the pressure to lower. The inversion weakens, and the dust will disperse, together with any fog or low cloud. Any convection which is able to take place will hasten the process. If the decaying anticyclone remains more or less stationary there will be little immediate change in the wind strength or direction, only in the appearance of the weather.

7.6

If the anticyclone is not decaying but drifting away, the wind will begin to increase and blow from a direction consistent with the changing location of the centre. The weather will also change because towards the edge of an anticyclone the inversion will be higher or weaker. Near the periphery of the system the wind may be strong, blowing perhaps up to 30 knots on the surface. It may, however, be less than this higher up. Smoke and fog will disperse and in summer cumulus will develop (unless the air is very dry) in a sky which is now of a brighter blue.

In practice, this means that if on a flight from London to the Midlands (Fig. 7.6) we have a NE wind on departure, the flight will be made towards the centre of the High. If it is a small one with its centre over the Midlands, disappearance of cumulus, slackening of the wind, and deteriorating visibility should be expected on the way. To the lee of, say, Birmingham, visibility might well drop below visual flight (VFR) limits. If, however, we fly in the opposite direction, towards Dover and Germany, and away from the high pressure centre, we will find freshening winds and improving visibility. Across the Channel, over Holland and eastwards, we will have every prospect of flying in the teeth of a strong easterly for some time – for which we should calculate sufficient fuel. Since the continental air is very likely to be dry there will probably be no cumulus. If we intend to fly above 4000–5000 ft, we should check strength and direction of the high level winds.

The fine weather produced by a small ridge of high pressure between two dismal depressions is always welcome, but it does not last long – often not more than 8–10 hours in Britain. The morning probably starts clear with a moderate westerly wind, which veers

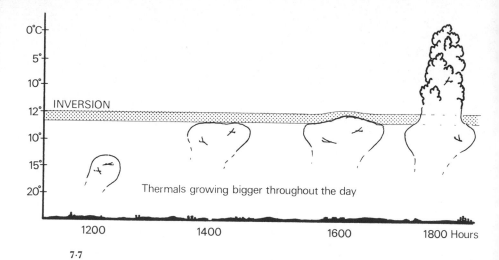

7·7

and slackens as the pressure rises. Cumulus may appear in the morning, and remain small or even disappear after an hour or so. As the 'crest' of the ridge passes, the wind may drop altogether for a short time, and then pick up again, at the same time backing slightly. As the wind freshens and backs further to, say, SW, frontal cirrus may become visible on the horizon.

Decline and fall

The decline of a cold anticyclone, in winter, may be first noticed when the sheet of dull stratus which has persisted for days, deteriorates into strato-cumulus, or lifts to the alto-cumulus level, and quietly evaporates as the lid above decays. If there is no cloud, the fog or haze will thin, and the amount of available daylight increase. In the grey conditions which have probably persisted for days, the first, almost imperceptible changes in the sky may easily go unnoticed.

In summer the anticyclone may decline or move away in the expected fashion, or it may be suddenly disrupted and broken down by violent thunderstorms. If an anticyclone has remained more or less static over land for some time, with a strong inversion at only a few thousand feet, the ground and the air below it will become increasingly hot. The warm air will expand and rise until it bumps up against the lid, still warmer at this height than the air that on rising has been cooling (Fig. 7.7). However, by the late afternoon

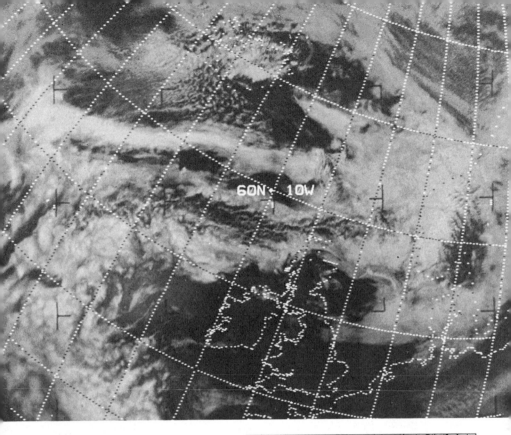

7.8 25 April 1972

The absence of cloud over Britain and
NW Europe is due to an anticyclone
centred, on the chart, just to the NW of
Ireland. Dry NE winds are flowing across
Germany and France with skies clear
except for cumulus too small to be visible
on the photograph. On this day a glider
flown by Hans Werner-Grosse beat the
World distance record by flying from
Lubeck to Biarritz, 1450 km in 12 hours.
Normally thermals are not available for
such a long period but by flying Westward
the day was effectively extended by an
hour.

On the satellite photograph the cold front
lying to the NW of the British Isles can
be clearly seen, including the greater
cloudiness at the W end where it is moving
WNW as a warm front. The cloudy tail
of another warm front over Scandinavia is
visible in the North Sea.

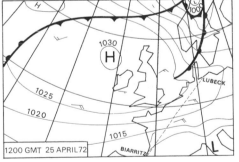

1200 GMT 25 APRIL 72

the temperature of the air just below the inversion is almost as warm, and the air near the surface really hot. Finally, some rising air becomes warmer than that of the inversion and breaks through. This is likely to happen in a number of places at much the same time. The warm air surging upwards through the inversion cools and forms clouds, which develop rapidly into big cu-nb thunderheads. Depending on the size of the anticyclone and on the terrain, the disruption may affect all or just a small part of the anticyclone.

Getting a forecast

When we can extract a lot of information from a weather map, recognise what we see in the sky, and relate this to the synoptic situation, we may feel that we can safely deal with poor weather flying. But if conditions are bad, or could turn sour, we should still try to find out more; so we go to the met. office. Since forecasters are not psychic we should help them give us the best information by telling them about the flight; not only where we are going, but whether the aeroplane is slow and what sort of fuel margin we have. We should also be honest about our experience, and not give the impression that we can fly competently on instruments for hours, when we cannot. If our airborne hours are low, or we are un-familiar with the aeroplane, this should be said.

Finally, we should go through the forecast systematically, and ask until we understand. A suggested weather briefing check list is given here, which can, of course, be altered or added to as required.

Weather briefing check list

GENERAL

1 What is the period for which the charts are valid?
2 What is the big pressure pattern, and which systems will influence our weather?
3 How are the pressure systems expected to move, and how fast are they moving?
4 What are the central pressures of the systems?
5 Where are any fronts located, and how are they moving?
6 Are the systems intensifying or decaying?
7 What are the high level winds doing?

ROUTE

1 What are wind speeds and directions at flying levels?
2 Is there likely to be precipitation, of rain, of snow?
3 What is the expected visibility?
4 What will be the height of cloud base, and of freezing level?
5 Is any marked deterioration likely at any point (local fog, heavier precipitation, extra low cloud base, etc.)?
6 Is strong convection likely, with thermal turbulence?
7 Will waves, or sea breezes develop, and if so, where?
8 Is there an Actual report available for the destination airfield, and for airfields to which it may become necessary to divert?

ADDITIONAL FOR GLIDING

1 What ground temperature is required before thermals will start?
2 How strong will thermals become, and with what cloud base at, say, 1300 hrs?
3 Is there an inversion, and at what height?
4 Is the instability such that cu–nbs may develop?
5 Is extensive spread–out of cumulus likely?
6 Is there likely to be any cirrus or other high cloud, and if so, how much?
7 Over what period are thermals likely to continue?
8 How far inland is any sea breeze penetration likely?
9 Where and when will any waves be likely, and to what height will they go?

Part 3   Special conditions

8 Fine weather cumulus.

The movement and behaviour of air on a large scale gives us the weather, but the sun shining on the ground also creates small scale convection in the surface layers. This mini-circulation produces up-currents – millions of them – and cumulus clouds, which modify the weather locally and sometimes substantially. Glider pilots know these upcurrents as thermals. The life cycle and appearance of the cumulus clouds they produce vary according to the characteristics of the prevailing air mass. They provide us, therefore, with a great deal of valuable information for both local and short term fore-casting for any sort of flying we want to do, so it is useful to get to know this family of clouds well.

Convection can take place only when sufficient heat reaches the surface to warm it, and the surface layers of air, to a higher tempera-ture than that of the surrounding air. Due to variation in the surface and the fact that any upcurrents have to be balanced by descending air, heating does not take place uniformly. When heated air at the surface becomes unstable it expands, rises and will continue to do so as long as it remains warmer than its surroundings. If, for example, the surrounding air is cooling at about the dry adiabatic lapse rate of 3°C per 1000 ft, and a bit of the surface and the air above it becomes 1°C or more warmer than this surrounding air, it will rise. It will not, however, continue to retain the 1°C bonus because, apart from any slight difference in the lapse rates of the two lots of air, there will be some mixing with the environment air which will tend to even out the temperature after a few thousand feet.

In winter, or in parts of the world where little warmth reaches the surface, upcurrents are weak, small – 200 ft or less across – and confined to just a few hundred feet above the ground. In summer, and generally in the tropics and sub-tropics, the thermal circulation may extend as high as 30,000 ft and dominate the weather during daytime. Upcurrents that start with a high temperature difference, perhaps 2–3°C above the surrounding air, may be a mile across, and rise at over 1000 fpm; the regular afternoon thunderstorms in the tropics being examples of what such thermals can achieve.

Convection is strongest at the time of maximum heating – in the

8.1 Cumulus

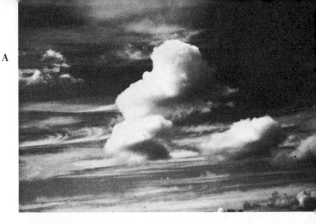

A Typical summer cumulus on a breezy day. The sparseness is due to the wind preventing sufficient warming of weaker thermal sources, and because the cirrus is reducing heating of the surface. This cirrus is indicative of strong winds at height, ahead of a depression.
July 4. 995 mb. Wind NW, 10 knots.

B Cumulus in moist but not very unstable air. The base is low for summer. The nearer cumulus is dark because it is in the shadow of another cloud.
July. 1700 hrs. Wind NW, light.

C and **D** Summer and winter cumulus. The summer cloud is firmer and crisper in appearance due to the continuing production of small water droplets in the rising air. The winter cloud, forming in much lower temperatures, has some ice crystals among the water droplets giving it a more feathery and insubstantial appearance.
July and November. Winds NW, 15 knots.

8.2

A A thermal grows from air warmed by the ground. This air becomes buoyant, expands and rises.

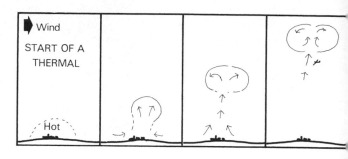

B Small thermals will be closer together than those which produce big cumulus.

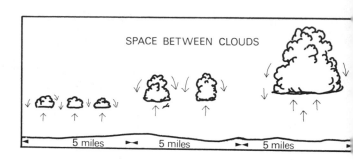

C The shadows from small cumulus do not affect thermal production, but shadows from large clouds, and also from high ground, will weaken it.

D Cumulus will sometimes lie in streets in strong winds. The long lines of cloud shadows help perpetuate streets by encouraging thermal production only in the sunny strips between the shadows.

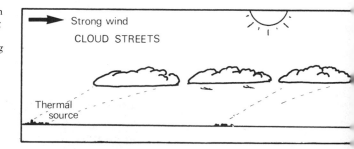

early afternoon – and weakest during the early morning.

There is negligible convection during the night, although up-currents may develop from warmth that has been stored during the day, as for example, in a deep forest. When the air has cooled down sufficiently, the now warmer air in the forest will start to ascend.

When a thermal begins its upward journey, air flows in towards the source to replace that which is rising, and this air is felt as wind. If the thermal is small and weak the wind will be barely noticeable, but if it is strong the gusts may be enough to shift a parked aircraft. Dust devils are visible signs of the departure of a powerful thermal.

On a windy day small weak thermals are easily broken up; even strong ones are affected, and to the pilot have a shredded feel about them until they have risen above the surface layer turbulence. If the strength of the wind increases with height, which is often the case, the thermal will slope downwind as well as drift with the wind.

Life of a thermal

Good thermals are born over ground which is dry and absorbs heat quickly, or which is surrounded by poor thermal-producing areas such as damp marshland. Towns and villages, corn or ploughed fields, desert, and large areas of concrete or factory roofing are all good. Poor areas include low lying soggy ground, otherwise good sources that have been rained on in the previous hour or so, and woods and forests early in the day. After some hours of sunshine woods will have had time to absorb sufficient warmth to produce thermals, and will often continue to give them later in the day, when more superficial sources have cooled off. In countries which are flat and sandy and liberally sprinkled with pine woods, such as Poland, we may find that the only good thermals on a day when the open country is kept cool by wind come from dry woodlands.

After a long period of drought, low lying damp areas which have become notorious for their lack of thermals, will have dried out and produce just as good upcurrents as anywhere else. However, soils which normally drain quicker, such as sand or chalk, are better than, say, clay.

Thermals are usually fewer and weaker along the coast, and may not develop at all when the wind is blowing in from the sea (Sea Breezes, page 158). If the ground is cool or damp, and the day windy, good thermals may grow from otherwise unlikely sources if

A

B

8.3 Birth and growth of cumulus

August. Wind NNW, light.
Elapsed time 26 minutes.
1744 hours. The first fragments
of the new cloud appear
overhead (A).
1745. The fragments beginning
to grow together as the full
strength of the thermal reaches
condensation level (B).
1749. The cloudlets are starting
to grow in size (C).
1752 (D). Looking like a good
cumulus. As it drifts away on the
wind another newer cumulus can
be seen at the top of the picture.
1755. Fully grown, with a soft
and fluffy top, and a flat, darker
base (E).
1758. Already decay is apparent,
the cloud is sagging, particularly
in the centre. The edges look less
soft (F).
1801. The cloud, now near the
bottom of the picture, is
flattening as it subsides (G).

C

D

E

1804 (H). The drooping, tired appearance indicates that no lift should be expected underneath.
1806. The remains are insignificant compared to the newer clouds now drifting overhead (I).
1810. Just a wisp left ($\frac{1}{4}$ in. above the bottom of picture). The later clouds are now all showing signs of decay (J).

F

G

I

I

J

they are sufficiently sheltered, such as large quarries, or shallow valleys. Sheltered SW-facing slopes, which have been warmed by even weak afternoon sun, may continue to give thermals for some time after they have died out elsewhere. Thermals may also be produced, or boosted, by some artificial source of warmth, such as stubble burning or houses on fire. Lift is good provided that the air is not too full of ash or oily smuts, which stick along the leading edge of the wing and tailplane. Having said all this, it is not really practicable in a glider to try to locate thermals by flying over likely sources, except when low down. Thermals initially climb slowly and so will drift a considerable distance on the wind before rising high enough to become of any real use. Generally, it is better to go by the clouds, or on 'blue' cloudless days, simply carry on along the intended track. Although usually of academic interest, the source of a thermal may sometimes be identified by its smell.

The lift-off of a strong thermal may not only cause gusts, but swirl dust, paper or straw into the air. This rubbish may be carried up several thousand feet, so that we can find ourselves circling with an old paper sack. Insects are swept up as well, and in turn attract such birds as swifts, which are prepared to go up several thousand feet for a meal. These birds do not use thermals for soaring, as do eagles or vultures, who will often circle with a glider, although they have been known to attack the intruder!

There are many theories as to the exact shape of a thermal, and we certainly find thermals of different shape in the air. Some seem to be like doughnuts, and others appear to be twin-celled, with two areas of lift very close to each other. The air in some thermals is extremely rough, but such thermals frequently have strong cores; others, usually large ones on quiet days, are very smooth. Thermals will give a rough ride to a fast aeroplane because it is rapidly traversing the alternate up and down flows; so it is better if we fly just above them in the smoother air. In gliders and slow aeroplanes thermals feel softer; ultra-lights can even make some use of them to save a bit of fuel. The energy that they supply is undoubted since they have supported gliders for distances of over 700 miles.

A thermal stops rising when its temperature becomes the same as that of the surrounding air, although if it has been rising rapidly, it will probably overshoot slightly, and then collapse back to its equilibrium height. If the air is dry, or the upcurrent is weak and has not risen very far, the thermal will remain cloudless. If, how-

B

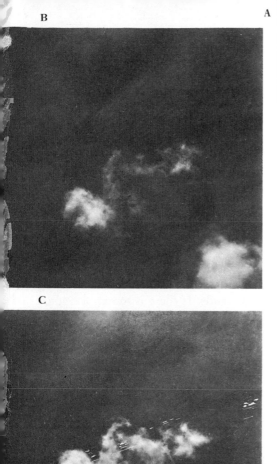

A

8.4 In dry air or under an inversion only the strongest part of a thermal may reach condensation level. This little cloudlet has appeared under a very strong inversion on an almost calm day. The patch of milkiness is caused by the condensation of some minute droplets. They may almost immediately evaporate again, or, if the thermal is strong enough, grow into more visible cloud.

C

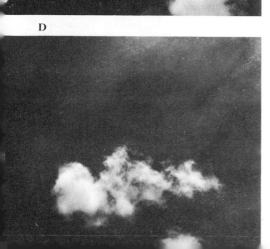

D

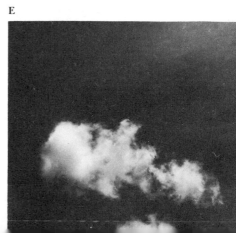

E

ever, it contains a normal amount of water vapour – in Britain on an ordinary fine day the humidity is somewhere about 70% – it will cool sufficiently on rising for condensation to take place. When this happens a cumulus cloud will appear. These clouds spaced over the sky on a summer day are excellent markers for the thermals which have caused them. They also indicate the height above which the air will be smooth.

The cumulus

On a day of good convection, condensation occurs at about the same height for all thermals, so cumulus clouds will also form at about the same level – condensation level. Looking at a sky full of cumulus we can easily see both the uniformity of cloud base, and the flattish base of the individual clouds. If there is some variation in the base it is because there is a difference in the moisture content of the air; for example, cloud base often lowers towards the sea coast. In very dry air, with only small amounts of cumulus, different sorts of thermal sources may well produce some variation in the cloud base, particularly early in the day. When the air is moist cloud base is normally even, but if there are some clouds that are giving showers, the base of the bigger storms will be lower than that of small cumulus at the same time.

When condensation takes place the creation of water droplets from water vapour generates heat – the latent heat of condensation. It is not much, but enough to give a boost to the upcurrent, which is then able to rise further and more strongly within the cloud. If the thermal is large, strong, and moist, the amount of heat made available will be enough to stimulate the cloud into rapid upward growth, with further condensation taking place from the remaining, cooling, water vapour. The growth rate may be enough to cause some of the surrounding air under the cloud to be sucked in as well as the thermal air, so that some self-stoking occurs. The water vapour in this air condenses, and the cloud grows rapidly.

When the warm air supply into the bottom of the cumulus ceases, the cloud will begin to decay away. The collapse is accelerated because as cloud evaporates heat is consumed in the process. The cooling cloud sinks further, more water droplets evaporate, and this process continues until the cloud has disappeared.

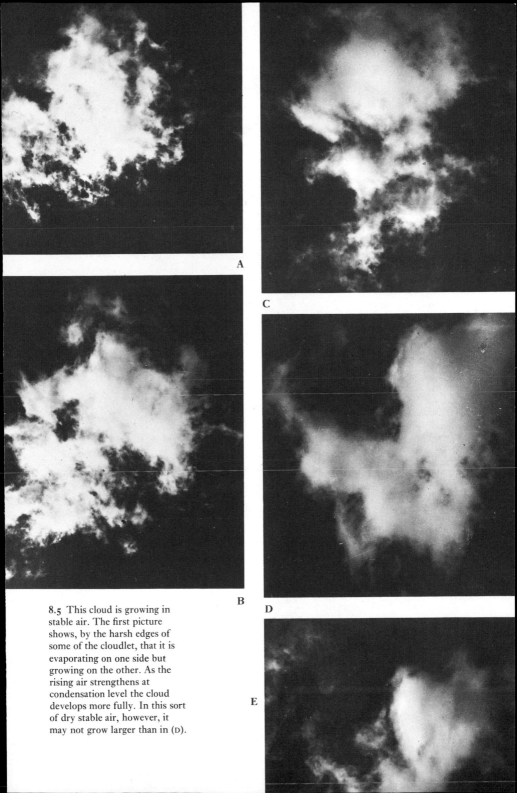

A

C

B

D

E

8.5 This cloud is growing in
stable air. The first picture
shows, by the harsh edges of
some of the cloudlet, that it is
evaporating on one side but
growing on the other. As the
rising air strengthens at
condensation level the cloud
develops more fully. In this sort
of dry stable air, however, it
may not grow larger than in (D).

A

June. 1100 hrs.

8.6 Cumulus growth

In light winds variation in the
strength of thermal sources is
more obvious than in fresher
winds, when cumulus have a
more even, cellular, distribution.
This is because the air remains
longer over the thermal source.
If a cumulus develops from a
particularly powerful source, it
may have a higher base, or grow
taller, or faster than its fellows.
The cumulus shown here
developed very quickly and went
on growing at the top at the same
time as drying out and decaying
at its base. The decapitated head
in the bottom photograph
remained floating higher than
surrounding clouds for several
minutes before dissolving.

The lightness of the wind is
apparent from the absence of
lean, or slope, in the cloud
structure.

B

C

Life of a cumulus

The simple summer cumulus (Fig. 8.3) has a short life, only lasting some 20 minutes between appearance and disappearance. This may be difficult to believe when we look at a sky dotted with puffy white clouds, as the impression is obtained that the same clouds float on and on across the sky. If, however, we watch an isolated small cloud it will be seen to dissolve and disappear quite quickly. In the meantime, elsewhere in the sky, new cumulus have been born. The new growing cloud can be recognised by its brightness and fluffiness, due to the high concentration of water droplets, whereas the dying cloud is more ragged in appearance and is faintly discoloured because the smallest droplets have evaporated; it is now composed of a smaller concentration of the larger ones, and the surface area available to reflect sunlight is reduced. The appearance of a cumulus alters not only because it is changing in itself, but because it is drifting along on the wind, and we see it from a constantly changing angle in relation to itself and the sun.

The time of day

Cumulus differ in their growth and appearance according to the time of day, due to the changing amount of available warmth. They also look different in spring, summer, and winter.

On a summer day in which we expect good cumulus to develop, the morning usually starts clear, or clears after mist or fog disperses. Initially the air will be smooth, discounting any turbulence solely due to the wind, but as the sun gains strength it will start to feel bubbly, and we find small areas of roughness up to about 1000 ft. If the air is moist tiny rags of cloud may appear at this height, and quickly dissolve again. The embryo thermals steadily improve, until by the time that they are rising to 1200 ft or so they will be almost large and strong enough to enable us to stay up. Sometimes all cloud disappears again for a time at this point because although the warmth of the sun is now sufficient to dry up the air near the ground, the thermals are still not going high enough to be cooled adiabatically below their saturation point.

On a day of very strong heating the lapse rate may exceed the dry adiabatic rate for a short time in the surface layers. This is because the ground, and air in contact with it, is rapidly warmed, and a little time is required to distribute this heat upwards. The reduction of

A

B

8.7 Cumulus growth

A remarkable cumulus which
developed quite uniquely at 0845
and vanished before 0900 hrs.
Nothing else like it appeared
that day (28 July). There was no
wind and the cumulus formed
above a south-facing chalk
slope of horseshoe shape, and
was probably caused by a pool of
warm air, maybe trapped in some
way overnight, and then warmed
sufficiently by the morning sun
to release it.

C

D

8.8 Cumulus throughout a day

Cumulus had started to appear at 0900 hrs and by 1100 hrs (A) were popping up all over the sky.

By 1200 hrs (B) they were growing in great profusion, and causing some cut-off of the sun. At 1500 hrs (C) the cumulus were large with big blue gaps through which the sun could continue to warm the ground. 1800 hrs (D). Thermals have ceased to rise from all except a few good sources. Cloud base is still high, but the cumulus is ragged and showing signs of collapse. June.

A

B

C

D

temperature will exceed the usual 3°C per 1000 ft in the first few hundred feet of height, but will return to normal as soon as the thermal circulation becomes established.

By about midday thermals are usually strong, rising at some 300–500 fpm in air which may feel quite rough. Unless the air is too dry genuine cumulus will now develop at the top of each thermal, probably at a height of about 3000–4000 ft; they will be compact, white, well separated from each other, and dotted fairly regularly across the sky. During the afternoon the clouds will become larger and somewhat more widely spaced. By about 1500 hrs cloud base will probably have risen as high as it is likely to go that day, maybe 5000–6000 ft, and the thermals will have reached their maximum size and strength, and they will have become less rough. As the sun declines, there will be fewer and fewer clouds as thermals cease to rise from the weaker sources (Fig. 8.8) during early evening. Cloud base will remain high but the strength of the thermals will deteriorate. Eventually, in the early evening the remaining cumulus will disappear and the air again become smooth. In hot countries thermal strengths may be up to 800 fpm by midday, with penetration up to some 8000 ft. In the afternoon lift may have increased to 1000 fpm with cloud base at 10,000–15,000 ft. Around latitude 30° thermals will continue to rise almost up to the time of darkness.

The distribution of thermals will vary with the depth of the convection layer. If this is very shallow, with thermals rising only to a few thousand feet, they will be relatively small and fairly close together. If, at the other extreme, thermals grow into cumulonimbus thunderheads rising to 20,000 ft or more, they will be more widely spaced. Early in the day, or when an inversion prevents thermals rising to more than, say, 2500 ft, they are likely to be about 200–300 yd across, with ordinarily not more than 1–2 miles between them. If cumulus cloud base is at about 8000 ft, normal thermals may be 700–1200 yd across and about 5 miles apart, but if cu-nb develop the storm centres may be 30 miles apart, with the storm itself up to 10 miles across.

Effect of the time of year

It is only to be expected that any cumulus appearing during the winter will be different from those of summer. Such cumulus as do develop will be at a lower height, more ragged, and of briefer

First 1000 km soaring flight

The first pilot to fly 1000 km (622 miles) in a glider, Al Parker, started from Odessa, Texas with the aim of flying northwards into Colorado to a goal 630 miles away. He started at 1029 hrs and landed $10\frac{1}{2}$ hours later. He circled in 43 thermals spending about 3 hrs 20 minutes doing so, and climbed an accumulated height of 69,2000 ft. The average rate of climb was 350 fpm, the average speed $61\frac{1}{2}$ mph, and the average tailwind component 18 mph. The release height was 5200 ft ASL, the greatest height achieved 14,200 ft ASL and the landing place 5150 ft ASL. The lowest point in the flight, at 1400 hrs, was only 700 ft above the ground. Late in the afternoon typical sub-tropical thunderstorms caused Parker to detour away from his goal; he finally landed at Kimball, Nebraska, 646 miles out.

The surface chart for the day shows the flight area to be between widely spaced warm and cold fronts. The curved isobars show that a slack low-pressure trough existed more or less along the flight line giving very light southerly winds, but that the upper flow, at 10,000 ft, was providing a stronger tailwind. The ground temperature in the afternoon was in the 90's in Texas and 100°F further up the route. Cloud base was 6000–7000 ft AGL during the morning, rising to 8000–10,000 ft later and developing into confused thunderstorms near the end of the flight.

The climate and topography of places like Texas are excellent for soaring. The latitude of 30°N provides hot direct sunlight, the air is dry and clean, and the ground dry and uniform on a fairly big scale. Little heat is reflected back off airborne dust particles, wasted drying up damp ground, or absorbed into forests or large areas of water.

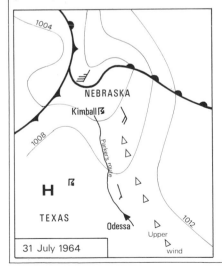

duration than their summer equivalent. In the spring the situation changes dramatically. The effect of the increasing power of the sun through the cold air and onto cold, but often dry, ground, sets off thermals that are rough and strong. Thermals remain good until late in the summer, although increasingly they will lack the vigorous qualities of springtime.

During the summer the daily convection period will be longer than in spring, with cloud base generally higher, and cumulus more widely spaced and often larger. As autumn approaches, the somewhat tired quality of the convection gives gentler and weaker thermals, which in due course decline away to ineffectiveness for the winter. The late summer and autumn weather in Britain and most of Europe can be excellent for flying training because it is often fine and mild for quite long periods. This is to a large extent due to lack of contrast; the sea and the land are both warm, and the power of the sun is declining.

Cumulus in moist air

In the cooler parts of the world there are many days when cumulus start to appear relatively early in the day – about 0930 or so – and the clear blue sky and bright little patches of cloud seem promising. If, however, the air is both moist and relatively stable with a slow temperature fall with height, clouds will increase in quantity fairly rapidly. The height at which cloud base occurs normally rises steadily throughout the morning, but on more stable days it will go up more slowly. As the cumulus grow, it is not long before they are at, or close to, the same temperature as the surrounding air. Unable to continue upwards, they will spread out and quite quickly the sky will become overcast as the flattening clouds meet. This reduces the amount of warmth reaching the surface, and so new thermals will be weak. Meanwhile those which created the cumulus will have died away so the clouds begin to collapse and evaporate; gaps or holes will appear in the cloud cover, the sun is again able to get through to stimulate the development of new thermals, and the cycle repeats itself. The periodicity of such cycles throughout the day, and the thickness of the cloud layer that is produced, will vary with the actual conditions on that day. For example, if the air is moist, the sun's heating strong, the air cool, and there is an inversion, say, at 5000 ft, the cumulus will develop and spread out over a thickness

A

8.10 Cumulus in dry and moist air

The top photograph (A) shows the largest cumulus that appeared on this dry air day. It is flat, with a smooth top where it is up against the inversion, and it lived only 4–5 minutes. The clouds (B) on moist air days are deep and domed, and if they do not develop too extensively cutting off the sun, they will grow to some 4000–5000 ft thick during the afternoon. Each cloud of this size lasts about 25–30 minutes.

B

of perhaps 3000 ft – from 2000 to 5000 ft. This mass of cloud will take longer to evaporate than when the air is drier, generally warmer, or if the inversion is lower.

If we intend to fly above the still separated morning cumulus on such a day, we should watch for signs of overdevelopment, as this will occur over large areas of sky often at precisely the same time. It is also possible that the 8/8 cover, once established, may not break until the late afternoon. Cloud base will remain at a safe height above low ground, so the main problem is only one of getting lost.

Such days are often unsatisfactory for cross-country soaring since the dead periods often result in the glider having to land prematurely. While sitting in a field waiting to be retrieved, it is infuriating to see the sky clear and new thermals and cumulus pop up all over the place.

Cumulus in drier air

When the air is very dry and the sun hot, thermals may reach their temperature equilibrium height without forming any cumulus. This does not happen frequently, the sky usually remaining blue because the thermals have come up against an inversion before reaching condensation level. Often in Britain the two are close in height. On such days we may find cumulus over ground which holds moisture, such as clay or green crops, but not over the better draining soils.

When there is a strong anticyclonic inversion at or just below condensation level, the presence of thermals may sometimes be detected only by milky-looking patches in the blue. A small amount of water vapour is condensing into minute droplets, and it remains suspended as long as the air in the particular thermal bubble is rising. When it finishes the embryo cloud subsides and immediately evaporates.

Thermals in drier air produce clouds that remain fairly flat, even plate-like, and well separated from each other (page 137). Because they are only small they will probably disappear earlier in the day than usual. On a day in which pressure is slowly rising, cumulus will tend to shrink during the day due to subsidence and warming of the air. Between the rising thermals and cumulus the air is descending, and this has a stabilising effect on cloud growth by reducing the lapse rate of the environment air.

Cumulus growing in mountains will mainly be found over the peaks, and not, as might be expected, over the warm valleys. This is because the rising air tends to flow up the side – the warmest side – of the mountains instead of free-rising from the middle of the valley floor. If cumulus has formed below the top of the peaks, which not infrequently happens, flying above a valley will provide a clearer, safer route.

Cumulus as a weather guide

Fine weather cumulus give us no problems. Flying above them is smooth, safe, and enjoyable; below them it will be bumpy, the visibility may be less good, and there will be more aircraft about – mostly gliders. If cumulus grow tall as towers in the morning, the afternoon could well be stormy, rough, and with heavy showers, if they flatten out under a clear sky and the lower levels are hazy, the weather is settling fine, if they flatten and disperse prematurely, the upper levels should be inspected for warm front cirrus, if they disappear on nearing the coast, the sea breeze may well have swung the wind around. While we float along we have time to look and to consider why the cumulus on this day are the shape and size that they are, and why their distribution is regular or patchy. When we are having to make up our minds as to what the weather is going to do, or not do, we obviously have to take information from all sources and feed it in to our mental computer – synoptic charts, barometer readings, the overall cloud picture, even the smell of the air. But we should not forget the cumulus. They are such good indicators that any information they can provide should be computed as well.

8.11 Cumulus in anticyclonic weather

In drier air, and under an inversion, cumulus remain flatter and usually well separated. Looking straight up at a newly-forming cloud (A). Where it appears milky further cloud may appear within a few seconds. 1100 hrs.

By 1300 hrs (B) thermals were strong enough to push up the bottom of the inversion, and form good small cu.

At 1500 hrs (C) cu reached their maximum size, and the sky had cleared by 1800 hrs.

May 20. Wind ESE, light.

A

B

C

8.12 Cumulus

A In light winds thermals may come off the same source every 5 minutes or so, but the cumulus they produce drift away so slowly that they appear as a procession.

The darker clouds are in the shadow of other cumulus.
July, 1000 hrs. Wind N, almost nil.

B A cumulus which has grown up into a decaying layer of old cloud, and is spreading out as it reaches it. The photograph was taken on a hill top, and the wind blowing up the face of the ridge had stimulated this cumulus, which obviously has orographic origins.
July. Wind W, 15–20 knots.

A

B

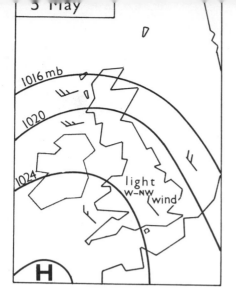

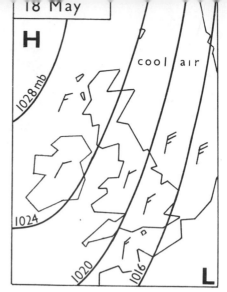

8.13 Four situations which give fine summer weather; good for flying and certainly good for soaring. In each case cloud base will be above 3000 ft and cumulus will remain small and well separated.

5 May. A high-pressure ridge, with warm air and light winds, but backing later.

18 May. Clean air in a light northerly flow.

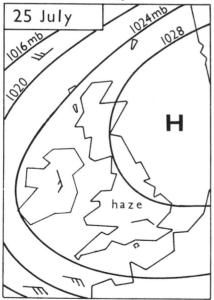

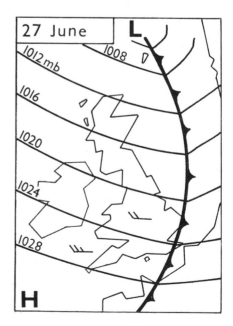

25 July. High over the North Sea bringing dry east winds from the Continent. Haze but probably little cloud.

27 June. Cool clean air after a cold front. Perhaps considerable cumulus but good visibility.

8.14 Cumulus distribution

On a day of uniformly fine
weather over a large land area –
in this case France – cumulus
sometimes develop in clusters
rather than in a random regular
pattern. This is particularly
likely over rolling upland country
such as the Massif Central or to a
lesser extent the English
downland. September 1971.

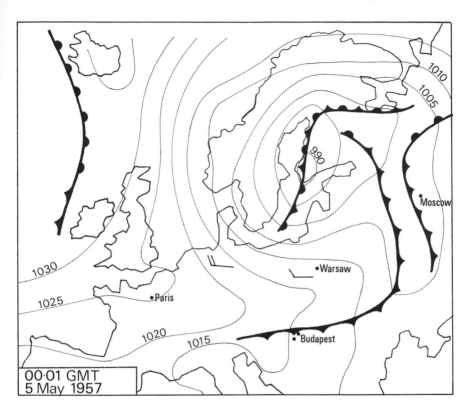

00·01 GMT
5 May 1957

8.15 The synoptic situation which gives the best chance of long
distance soaring across Poland. Many flights of 500 km were made
from Leszno, in the west, to the Russian frontier not only by Poles,
but by visiting glider pilots from all over the world. I flew this
route in 1961, in a Jaskolka, taking 8 hr 5 min for 528 km, landing
at my goal of Tyszowce. Cloud base was 6000 ft, with a thermal
strength of 6–8 knots in the early afternoon. The west wind was
15–18 knots at 2000 ft.

9 Cumulo-nimbus thunderheads

On some days cumulus will continue growing larger and taller until they become huge cumulo-nimbus thunderheads. For this to happen conditions have to be unstable. Instability, in this sense, means that when air starts rising it will continue doing so because there is no barrier of warmer air to stop it. Eventually, of course, the thermal and its cloud will cease rising and growing when warmth ceases to be fed in at the bottom.

The conditions needed to create a highly unstable situation are:

(1) that the air shall cool with height at the dry adiabatic lapse rate, uninterrupted by significant isothermal layers or inversions – if the lapse rate is initially higher than the dry adiabatic rate just above the surface, so much the better;

(2) the air mass shall be reasonably moist; and

(3) the surface heating shall be strong.

The sky on such a day looks clean; thermals often start early, develop vigorously, and rise rapidly. As condensation level is reached cloud appears and it continues to grow upwards. As it increases in size the surrounding air is pulled in from below by the powerful updraught and boosts the cloud. Sometimes the cloud growth is so rapid that it pushes up the clear air above it fast enough to cause this air, in turn, to produce cloud. These clouds look like a cap or an eyebrow and are called Pileus, meaning 'a felt cap without a brim worn by the ancient Greeks and Romans.' If one is seen, the cloud head beneath it could be rising at over 1500 fpm (Fig. 9.2).

Once a cumulus has started to consume substantial quantities of air in addition to that of its own original thermal, it not only grows upwards but increases enormously in volume, becoming composed of a number of rising cells, rather than a single upcurrent; sometimes cumulus heads in a developing cumulo-nimbus can be seen growing while other parts of the cloud are visibly collapsing.

When the cloud rises above freezing level some ice crystals will form among the water droplets; the proportion increasing as the rapidly rising cloud penetrates the lower temperatures of the upper air. By −40°C all the particles will have frozen. When much of the

9.1 A day on which thunder-storms are likely to develop starts with smaller cumulus, but often they grow quickly tall for their size. (A) and (B) were taken from 28,000 ft at 1000 and 1200 hrs respectively, over Germany in August. (C) shows a rapidly growing cu–nb with the ice crystal anvil leaned over by strong upper winds. Texas, 1300 hrs, July.

A

B

C

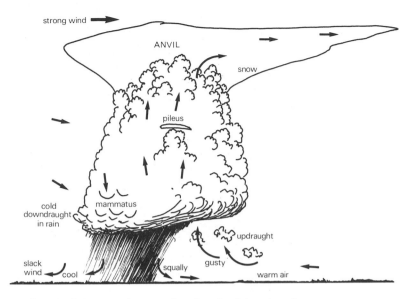

strong wind

ANVIL

snow

pileus

cold
downdraught
in rain

mammatus

updraught

slack
wind
cool

squally

gusty

warm air

9.2 Cumulo-nimbus grow from cumulus when the air is moist and
unstable, and the sun hot.
They develop a powerful circulation of their own and travel
across country slightly faster than the ordinary wind. The top of
the cloud is formed of ice crystals, the inside may be very turbulent
with risk of icing and lightning strikes, and the rain or hail falling
from the base can be very heavy. The appearance of small pileus
clouds shows where the upcurrents are very strong.

cloud has become composed of ice crystals the cauliflower, or mush-
room, shape of the typical cumulus gives way to the characteristic
anvil formation of the cu–nb. In extremely cold weather, or if the
cloud starts to form only at a great height, it may be made almost
entirely of ice crystals, in which case only the wispy anvil cloud will
appear. Even if there is some ordinary cumulus this may disappear
before the anvil, since ice crystal clouds evaporate much more
slowly than water droplet clouds, and it may last well into the night.
If there are strong winds at high levels the anvil will stream downwind, sometimes looking similar to a feathery finger of dense frontal
cirrus.

All the time the cloud is growing an increasing number of ice
crystals are sinking down through it, accumulating moisture, and
growing bigger. As long as they continue to fall through cloud
which is itself rising faster they will go on growing. In due course,
part or all of the cloud will cease rising, and the droplets, now quite

Cumulo-nimbus thunderheads / 147

9.3 Cumulo-nimbus growth

When a cumulo-nimbus cloud grows upwards very fast, it also pushes up the clear air above it. Cloud sometimes develops in this raised air, and this can be seen happening here (A). Often Pileus cloud looks like a cap above the big cloud.

When a cumulus grows above freezing level ice crystals begin to form and in due course parts of the top of the cloud are made of ice crystals and the cloud develops an anvil-like appearance (B and C). Inside these clouds the air can be really rough with severe icing probable. Cloud base may be lower than for ordinary cu at the same time, and with rain or hail falling from it. The surface wind will fluctuate in direction and is likely to be very gusty.

A

B

C

July 7, 1030–1100 hrs. Wind W, light.

A

9.4 Cumulo-nimbus

In very cold air cu-nb may form
almost entirely of ice crystals,
with little or no water-droplet
cumulus at the bottom.

This has occurred (A) and the
anvil is now rapidly decaying.

Ice crystal clouds are colder
and longer lasting (B). Here no
cumulus has developed at all.

Heavy rain falling from a big
cumulus with the sun shining on
it (C). Away from the sun, the
rain is in shadow.
15 April, 1720 hrs.
Wind NW, 12–15 knots.

B

C

large, will fall out as heavy rain. In winter the fall may be of snow which sometimes melts before reaching the ground; or it may even melt, and re-freeze, while passing through a cold layer of air, and reach the ground as frozen rain. Since a small cu-nb, or cells in a large one, will last for rather less than an hour, the rain will be of short duration. However its intensity will make up for this deficiency, since rain from a cu-nb may be between 10 and 100 mm/hr. The loss of this moisture from the cloud may increase its buoyancy so that it surges up again, but more usually once the heavy rain has started, the cloud starts to decay.

Some cu-nb will produce hail. The fall may be slight and mixed with rain, or it may be a diabolical bombardment of lumps of ice the size of golf balls, or larger. Since damage done by hail may be catastrophic it is useful to be able to predict what is going to lash out of any ominous black mass that is coming our way. As we have seen, if hail is to grow large, the ice crystals have to fall through very high or cold cloud long enough to accumulate the millions of tiny droplets necessary, and they can only do this if the cloud through which they are falling is also carrying them up, or at least supporting them for half an hour or more.

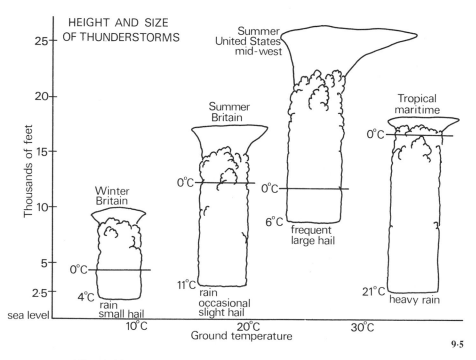

9.5

An interesting thing about thunderstorms is why in some places nearly every storm will produce large hail, and in other places equally big cu-nb never have hail at all. For example, moist tropical thunderstorms just produce very heavy rain, but storms growing in the dry air of say, the American Midwest produce devastating hail most of the time. The reason appears to be associated with the proportion of cloud above and below freezing level (Fig. 9.5). Over the sub-tropical regions of Texas and New Mexico a situation exists which may lead in summer to the development of super storms; this area is deeply penetrated by hot moist air from the Gulf of Mexico, which lies adjacent to the equally hot, but dry, air of the desert regions. The shear line between the two may be almost unnoticeable or it may separate quite different weather. Sometimes the shear line slopes, or moist air is trapped underneath a mass of drier but cooler air, and a potentially highly unstable situation then exists. All that is required is for strong thermal convection to create enough of a disturbance to release the hot humid air. This will burst upwards with great intensity and power, and grow rapidly into a large and really violent storm. The fact that it may not have been possible to forecast accurately the likelihood of such an occurrence or at any rate its location, is of no help if we are at the receiving end of the tornado strength winds or pigeon's-egg hail. The only consolation is that these super-storms result from conditions peculiar to a region, and have happened before; so we should accept as deserving respect *any* local warnings when aviating in such areas.

Flying in, or near, cu-nb should not be undertaken lightly, even in quite big aeroplanes and even in mild weather countries like Britain, where really fierce storms are infrequent. Apart from turbulence, one of the risks in damp countries is the lowness of cloud base. In mountainous regions the storm base may frequently be lower than the peaks. The bottoms of cu-nb are frequently lower than those of ordinary cumulus on the same day, with visibility below cloud remarkably poor due to snow, hail or rain. The precipitation comes from the least active part of the cloud, usually the back, with the largest raindrops falling at about 8 metres a second. The biggest downdraughts will be found in the rain, particularly where it is evaporating and further cooling the air on the way down. In a big storm this cooling may be enough to produce a local pressure rise equivalent to about 200 ft on an altimeter. Since the approach of the storm will have previously given a gentle, local pressure fall – making the altimeter over-read – not too much faith should be placed in this instrument when flying close to big storms and needing to know the height accurately.

Perhaps the biggest risk with cu-nb is getting involved with them by mistake; becoming iced up so that the engine or airframe performance is impaired, and then finding that the combination of low cloud base and precipitation have made escape, or a successful emergency landing, just too difficult. Another risk, if flying in or close to cu-nb, is of damage from hail or lightning strikes. Hail can alter the wing leading edge profile to such an extent that the stall speed is substantially increased, even if it does not actually puncture the wing skin. With lightning strikes an unbonded aircraft may suffer control or structural failure (page 157). In countries where thunderstorms are known to grow to great heights and severity, there is no question that the best policy is to stay as far away from them as is possible. The cloud is, after all, only the visible part of an extremely disturbed mass of air. On the downwind side, particularly, severe turbulence may be found outside the cloud, or in wisps of anvil cloud, even as much as 20 miles away. Above a very high storm the wind may be blowing at 100 mph, perhaps more. If we want a quiet ride we should keep clear of the cloud by some 10,000 ft above its top, although this is an academic point since most light aeroplanes have an effective ceiling about half way up such a storm. Certainly we should not attempt to go through any slots under big storms or line squalls. Inside, the up and down-draughts may exceed 3500 fpm, and have caused structural failure of aircraft flying into the narrow but extremely turbulent transition zones between the vertical blasts. Upcurrents of these strengths are powerful enough to support a parachute inside cloud for half-an-hour or more, which will probably result in the unfortunate pilot being frozen, starved of oxygen, or charred by lightning. If the pilot delays opening his chute to avoid these hazards, he risks hitting the ground through failing to see it in time in the heavy rain or hail; rain drops hitting the eyes at about 60 mph make it virtually impossible to see. This is not theory; these things have happened.

If forced to land in a field by a big thunderstorm, our first action is to secure the aircraft. This will be urgent if we have landed just ahead of the storm, otherwise it will be turned over by the violent gusts blowing out from the cold air downfall like a fierce cold front (Fig. 9.2). The wind comes with surprising suddenness, often following light airs which have encouraged a false sense of security. When tying down, or getting the aircraft behind a windbreak, it should be remembered that the direction of the gust front will be associated with the direction of movement of the storm, which may not be the same as that of the general wind. If we are caught out far from habitation some sort of personal shelter from the rain – and hail – should be found. If there is nothing, a garment should be taken off and kept dry. Although the ground temperature before

---

Paris to nearly Nice

Due to reported low cloud in the high Mediterranean Alps we chose to fly Lyon, Avignon, Nice direct. Everything proceeded according to plan until we saw ahead large and magnificent cumulo-nimbus, 5/8–6/8. These provided some interesting flying as it was easily possible to fly between the individual cu-nb. However, on approaching the coast the cloud density increased and gaps were difficult to find.

So near and yet so far! We had only fifteen minutes flying time left to Nice, and decided that the icing and turbulence would be for a short time only. Entering cloud at 8000 ft produced two immediate results. Firstly, severe airframe icing with build-up about $\frac{1}{2}$ inch per minute, and loss of ASI (we forgot *Pitot heat*). Secondly, severe turbulence. A rapid decision was taken and the aircraft put smartly about on a heading of 270° mag.; then a couple of circles while we decided the next course of action, and we steered north toward the high ground (up to 15,000 ft) to attempt to circumnavigate the cu-nb. Unfortunately they stretched as far as the eye could see – visibility was good. Two minutes feverish VOR/ADF plotting followed, which put us 10 miles north of Cannes. Ten nautical miles at 150 mph equals 4 minutes flying – it seemed so little. Due south was selected and we entered cloud at 9000 ft which gave 2500 ft terrain clearance.

Thirty seconds later the aircraft was bouncing along, the vertical speed indicator was off the clock and the noise of the hailstones was deafening. Outside air temperature was minus 10°C, cockpit temperature boiling. The next three hours (actually 8 minutes) were very exciting. Nothing changed except the colour of my hair and the leading edge paint work – until miraculously directly ahead was Cannes airfield. Blue sky, calm sea and behind were the wild and angry cu-nb sitting on the mountains. As far as I am concerned, never again.

—Account by D. Adlington

9.6 Cumulo-nimbus

A big cumulus developing into a cumulo-nimbus. Behind is the old anvil of an earlier cu-nb, the rest of which has decayed away (A).

In (B) a small squall line is producing cu-nb heads. On the right a new ice-crystal anvil head is growing. A few pileus clouds show where upcurrents are strong. The decaying cloud overhead is the remains of an earlier big cumulus which has collapsed and is evaporating.

9.7 Cumulo-nimbus in winter

(A) A small but powerful cu–nb in the polar air after a cold front. The low base, and the lack of much cumulus below the ice-crystal anvil are typical of winter storms. The cloud is moving away from the observer. In (B) the sun shining on the rain, and the mammatus bulges can be seen. (C) shows a late decaying stage of the cloud. All cloud disappeared within the hour and the night was clear and cold.
25 October, 1645, 1710 and 1720 hrs. Wind W, 20 knots.

A

C   B

the storm may have been 25–30°C, the rain will have only just thawed, having started as falling snow or hail at a temperature of −40°C or less. If we are thoroughly soaked by such cold rain and it remains cloudy after the rain has stopped the experience is chilling and continued evaporation of both ourselves and the ground will keep it that way for some time. If big hail starts coming down it is *essential* to find shelter, even at the expense of the aircraft. In the States, pilots running for cover have been beaten unconscious by hail, and then blown along the ground in the wind through prickly pears and cactus.

The only good thing about hail is that its duration is usually brief. The cell of the cloud which has produced the icy chicken eggs, golf balls, or whatever, boots them out as soon as they become too big to be supported any longer by the cloud, and that is it. The cloud cell decays away, and further hail must come from another, still growing cell. By the time that this hail is ready to bombard out, different countryside and other unfortunate people will be underneath.

Because summer thunderstorms go to much greater heights than winter or cold front storms, their hail will generally be larger and harder. The largest hail of all comes, as we have seen, from subtropical thunderstorms over hot, dry, plains. Various authorities have tried to categorise hailstones so that people can give reasonably accurate information on their size, if they do not have a freezer handy. Most graduations start with grain or shot, increasing through currant, pea, grape, walnut and golf ball, up to tennis ball; but all this can only be an approximation since a Texan pea is probably bigger than an Alaskan one.

The pilot said . . .

. . . he had been flying at 9000 ft – attempting to remain VFR on a cross country flight – when his aircraft was sucked into a thunderstorm. Severe turbulence and hail tore away large pieces of wing fabric and bent the left wing downwards. The Stinson then spun until it broke cloud about 2000 ft above ground. At that point the pilot said he regained partial control and crash landed in an open field.

Loose electricity

Lightning occurs between a cloud and the ground, in cloud, or between two clouds. The greatest risk is within about 5000 ft either side of freezing level, but lightning will also strike outside this region. The effect of lightning is variable, and generally on an all-metal aircraft insignificant. On wood and glass-fibre aircraft the presence of the metal parts may have odd effects – a control cable may be burnt through while the wooden wing in which it is contained appears to be virtually undamaged. Alternatively, a strike passing through wood or glass fibre on to the metal beneath may reduce the non-metallic parts to the consistency of pith. Sometimes a pin hole in the surface may be the only external sign that there is trouble inside.

A lightning strike may cause the aircraft's ferrous parts to be magnetised, or the compass magnetism to be upset; either may result in the compass giving misleading readings.

Aerotowing of gliders in thundery conditions is usually merely unpleasant, because of the turbulence which is encountered; however there is possibility of a higher than usual strike risk since the tug, tow rope and glider are, in effect, one big object – although there is no evidence of accidents from this cause. Winch-launching is obviously very hazardous, since the wire cable sticking up 1000 ft or more into the air becomes an excellent lightning conductor. On some gliders with poor electrical bonding the pilot's body will be connected to the aircraft by his right hand on the stick, and to the ground by his left hand on the release knob. A quite small potential gradient becomes appreciable when extended vertically through 1000 ft; even if a strike does not occur there is still a risk of the pilot getting a shock.

Most cumulus clouds are produced by separate thermals occurring within a homogeneous layer of air. There are, however, many occasions in which masses of air having different characteristics come together. We have seen this on a vast scale in the circulation of depressions, and on the frontier, or shear line, between, for example, air from the desert and the ocean. If the temperature is similar the two lots of air will tend to remain adjacent but separate. An example is the so-called Marfa Dewline in SW Texas, where the warm dry desert air meets warm moist air from the Gulf of Mexico, and it becomes sufficiently established to be regarded as a semi-permanent summer feature.

Sometimes, as when the sea air has been over land for a while and has dried out, it is difficult to tell where the junction lies because there is little difference between the two lots of air; there may be just a change in visibility from clear air to a slight haziness. If there is an appreciable difference in the humidity there may be little or no cumulus on the dry side, and big cumulus or cu-nb on the moist side. In due course, this difference in the extent of the convection will result in pressure differences, and the two lots of air will start to react on each other. If there is some convergence motion air will be pushed up along the dividing line, and if condensation level is then reached cloud will form along the line. If the reason for the convergence zone is topographical, as between land and sea, the size and location of the zone will be related to the surface features which caused it.

In all maritime countries there are frequent meetings, or convergence, between incoming sea air and the air over the land, particularly in summer. This creates a wind, known as a sea breeze, which may appreciably affect the local weather. If there is strong convection with large quantities of air rising over the land, air from the sea will be drawn in to even out the pressure discrepancy. The sea air is usually cooler, and it will penetrate inland increasingly throughout the day, perhaps as far as 20–30 miles, until convection weakens with the decline of the sun.

The sea wind will then drop, and as the land cools extensively during the evening and night the flow, if any, will reverse. In the

10.1

tropics sea breezes provide a valuable amelioration to the heat, although on beaches in temperate climes they frequently only create a sand-blasting machine.

The behaviour and appearance of the sea breeze, and the distance inland that it will penetrate will depend on the angle at which it meets the main wind, the direction of this wind in relation to the coast line, the extent of the convection inland, and the temperature of the sea. Of these, the factor likely to make the biggest difference to coastal weather is the directional relationship between the sea and main winds (Fig. 10.1).

Let us assume that the day has started fine and warm. At perhaps 0930–1000 hrs, cumulus can be seen to be developing some miles inland. The main wind is light to moderate and, say, easterly in direction (A). On windward coasts, in this case the east coast, sea air will penetrate quickly and easily, since it is simply augmenting the main wind. Any cumulus which formed early near the coast will disappear, and further clouds are unlikely to develop in the region of the east coast, so the day will remain bright and sunny though cool. It is not actually impossible for a few small cumulus to grow within the sea air if the wind is light, in sheltered places, or where the sea air layer is shallow and there is high ground above its influence; warm air will start rising from the 'island'. The base of any cumulus forming within the sea air region will be lower, and the cloud less well shaped, than those developing in the drier inland air.

On lee coasts, in this case west-facing ones, the sea breeze will be opposed to the main easterly wind, so its penetration will take place more slowly and probably go less far (B). As soon as convection gets going over the land, air from the sea will move in, although at this early stage the wind at cloud height will not have changed; initially the sea breeze will affect only the surface levels. Along the frontier between the two winds there may well be a narrow belt with no

Cloud streets and convergence cumulus / 159

wind at all. Cumulus which have formed inland may, of course, be seen to drift out over the coast, dispersing a short distance out to sea. On the coast, in the early morning calm, it may well both be, and seem, warmer than it does later after the sea wind has started to bore in.

As the sea breeze strengthens and penetrates further, cumulus which had been able to form only a short distance inland from the coast will disappear, and convection will retreat steadily inland. In the hinterland big cumulus may be visible, sometimes looking like a continuous wall of clouds to the observer by the sea.

So far we have looked at simple situations where the main wind either directly augments or opposes the sea wind, but most coast-lines are irregular and alter in direction. More often the two winds will converge at some angle to each other. Along the line or zone of convergence there will be some upthrusting of air as the cool moist air undercuts the warm land air. When this happens, cumulus will develop along the convergence zone forming a line or belt of cloud; the amount will vary according to the degree of convergence imposed by the coastal shape. This is known as a sea breeze front (Figs. 10.2 and 10.3).

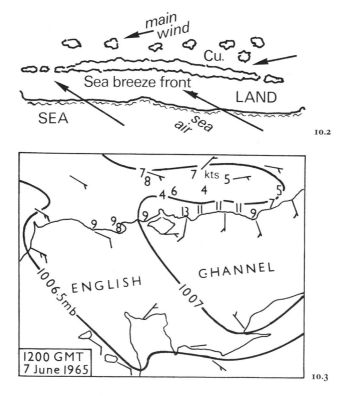

10.2

10.3

10.4 Coastal cumulus

Locally winds near the surface may converge or meet due to ground obstructions or due to temperature difference caused by surface change; e.g. land and water. (A) shows a line of cumulus cloud which developed to the lee of the Start Point peninsula in the early morning. The wind was light, westerly, and the cloud developed on a line where the air flowing up Channel over the warm sea coincided with the night-cooled air drifting off the land. As soon as the sun got high enough to warm the land (about half an hour later) the line of cloud dispersed. The effect of the convergence was enough to cause the light showers which can be seen under the cloud. August. 0815 hrs.

On days when the land is well warmed by the sun, the rising air over it pulls in air from the cooler sea. Even in late summer when the sea reaches its highest temperature, there may be a difference in the temperature of the two lots of air. The air from the sea is pulled in generally towards the mass of the land, and what happens when it meets the main weather pattern wind depends on the relative angles of the two flows. In (B) a sea breeze convergence zone has developed along a NE–SW coastline between a NNW main wind and an easterly sea breeze. There is usually some penetration of the sea air during the course of the afternoon, but this one remained stationary a mile or so inland of the general line of the coast for several hours. August. 1500 hrs.

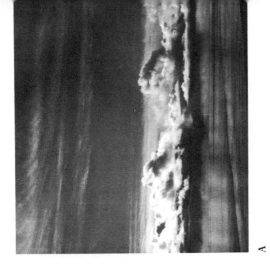

A

B

As would be expected, cloud base along this zone will be lower than that of the prevailing inland cumulus. Flying towards the coast underneath cloud, the lower cloud of the convergence zone sometimes gives the appearance of a complete clamp ahead. If we do not wish to risk a headlong penetration, flying parallel to the coast should soon produce a gap, particularly where the coastline is irregular in shape. The sea breeze front may lie all day only a few miles in from the coast over the same ground, or it may steadily penetrate further inland, depending on the relative strength of the two winds; but it will lie roughly parallel to the general coastline. The front will develop most strongly in the afternoon, at the time of maximum heating, and it may persist into the evening. In regions where sea breeze fronts occur fairly regularly, such as along the south-facing Dorset coast, there will be a noticeable difference in the annual amount of sunshine received by places only a few miles apart. The line of rising air which is found along an active sea breeze convergence zone often provides lift in which a glider can be flown more or less straight for considerable distances without losing height.

A convergence zone may be present without cloud if the air is very dry; in such cases all that may be noticed is a small change in visibility or in the colour of the sky due to the different moisture or dust content of the two lots of air.

As we would expect, sea air will penetrate more readily where the land is flat than where it has to climb up over hills, certainly initially. Thus the sea breeze is likely to be held back where there are cliffs backed up by high land; but since the reason the air is being pulled inland is to even out the pressure fall due to convection, it will have to get there somehow. Estuaries, fjords, and wide valleys will therefore receive extensive penetration of sea air on days of strong convection and an appropriate main wind direction. If the air is very moist, and there is a strong wind assisting the penetration, the valley or fjord may fill up with cloud, which may be very low or even on the ground.

Over peninsulas sea breezes become complex. Early in the day, as soon as convection get going, penetration of sea air will begin on all coasts, but what happens thereafter will be controlled largely by the prevailing wind. If this is straight down the peninsula, or if there is no wind so that convection alone dominates the situation, a convergence zone may develop like a ridge along the spine (Fig. 10.5). More usually, however, the wind tends to blow across the

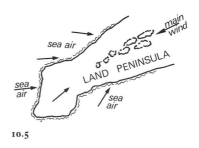

10.5

tongue of land which becomes largely submerged in the sea air, and any cloud that has already formed will probably disappear. It goes without saying that the surface wind at the coast should be checked before landing – not only for direction but for strength.

The effect of cloud shadows

As soon as cumulus appear, the surface heating will be modified by the passage of their shadows over the ground; this in turn will affect the growth of subsequent cumulus. If there are only a few clouds in the sky the effect will be negligible, but if a considerable proportion of the sky is covered, the heat from the sun will be reduced over large areas of ground. The situation resolves itself after a while because the cooled ground will produce fewer thermals, and the existing cloud will decay away. The sun will then be able to warm the surface again and new thermals will develop. On such days there is often a noticeable periodicity of alternate cloudy and sunny spells; in the British summer the overcast periods sometimes occur at quite regular 2-hour intervals.

If there is a wind blowing, the thermals and their cumulus may lie in lines, called cloud streets, usually about parallel to the wind. The clouds throw long lines of shadows which confine the sunshine to the ground between them, and encourage thermals to develop only in these slots. This self-perpetuating feature of cloud streets is, however, often frustrated by other factors, such as the difference in wind direction between the surface and that at cloud height, and the changing angle of the sun to the line of the streets – as well as by the normal processes of cloud decay.

Cloud streets may also appear in very light winds simply as a result of thermals coming successively off the same source; before one has drifted very far, another one has popped up.

Soaring and sea breezes

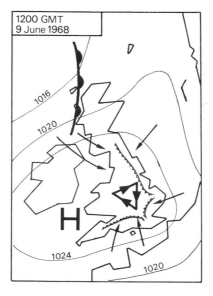

1200 GMT
9 June 1968

10.6

Extensive sea breeze penetration occurs when the weather over Britain is fine, warm, and with light winds. Such a day was 9 June 1968, with an anticyclone centred over Britain. There was some broken layer cloud in the North, but 10–13 hours sunshine everywhere else. Sea breezes started moving in along the South Coast at about 1000 hrs, and blew steadily all day at about 10 knots from 200° (SSW). The general anticyclonic wind was less than this strength and from the NE. The convergence of these two winds produced a line of cumulus – a sea breeze front – which moved steadily inland at 2–3 knots. It passed over Lasham, Hampshire, just after 1700 hrs pretty well packed full of soaring gliders, penetrating to Reading at 2040 hrs still surprisingly active for such a late hour. Even as far inland as Abingdon it was traceable at 2200 hrs, sunset.

There was extensive sea breeze penetration along the east coasts, but generally no typical convergence zone since the sea air and main wind were too similar in direction, and merely augmented each other. However, further north, where the anticyclonic wind had pulled round to a very light westerly, a soarable convergence zone was more likely.

The first 500 km triangle ever to be flown by a glider in Britain was made on this day (Brennig James, Diamant 18), using thermals inland of the sea air. His main problem was, in fact haze and sea fog which weakened the lift brought in by the easterly flow from the Wash. The course of the flight is shown.

A

10.7 Cloud streets

Typical cloud street forming in
moist air and a fresh wind. Cloud
base is low, only 2700 ft (**A**).
(**B**) cloud streets under a strong
anticyclonic inversion. These thin
flat cumulus were the only clouds
to appear for several days – from
about 1000 hrs to 1700 hrs each
day.

The streets were over clay. No
cloud appeared over the even
drier, adjacent sand and chalk
areas.
September. 1430 hrs. Wind ENE,
18 knots.

B

(**C**) During the afternoon of an unstable day with cu-nb
development and heavy showers, the air became less unstable, and
cloud streets started appearing instead of the storms.
16 June. 1500 hrs. Wind SE, light to moderate.

C

10.8 Cloud streets

Most cloud streets lie with the wind, but these were different. They were across wind.

The sky had been almost overcast for some hours, and broke during the late afternoon.

As the sun sank below the edge of the higher cloud sheet and shone on the ground, these rolls of strato-cumulus appeared.

No explanation is offered. They lay over a valley near to a range of hills, to which the wind was blowing parallel.

The clouds dissolved as they drifted away and were gone in about 10 minutes.

April. 1700 hrs. Wind NE, strong.

B

C

If the surface covered by a cloud shadow has time to cool appreciably in relation to the surrounding sunlit ground, as it may if the cloud is large and the wind light, the drifting of the shadow over the ground creates a travelling pool of denser air pushing along the surface. This may be enough to stimulate embryo thermals to lift off just ahead of the shadow, before they would otherwise be ready, so they may not go high enough to produce cloud. It is noticeable on the ground that when the shadow of a big cumulus arrives the air not only feels suddenly colder than expected, but the wind temporarily increases in gusts.

Cumulus distribution

Although superficially the appearance of cumulus appears to be an entirely random affair, on studying any given locality it is surprising, when wind and convection are similar, how often they appear in the same place. Convergence zone cloud banks are the most obvious example, but anywhere that the ground is not flat and uniform there will be some reason for the cumulus pattern. Normally this is unimportant either for flying or living, but in coastal areas where airfields can quickly become swamped by cloud, or high-level airfields where cloud base may often be only marginally higher than the landing place, a study of cumulus distribution in different conditions, and particularly in moist air, will provide a lot of information on local cloud development patterns and weather.

11 Lenticular and other wave clouds

We know that when air is confronted by a hill or mountain massif it will rise over the top in order to continue on its way. In fact, some air will flow around the ends of a hill or short ridge, but the main movement of the air is an ascending one; the slope-soaring carried out by gliders is simply a matter of flying continuously in the region

---

New Zealand waves

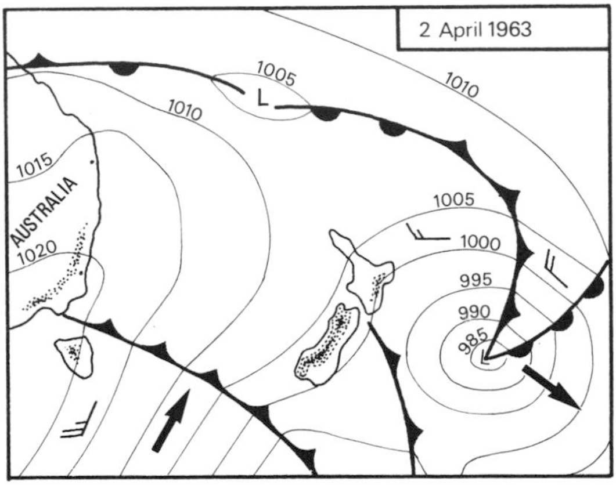

2 April 1963

11.1

New Zealand is an ideal place for waves. Its high mountains form a gigantic spinal ridge lying in the strong wind belt of the Roaring Forties. Because the mountains are surrounded by sea and only a small strip of flat land, the airflow is relatively uninterrupted, particularly by the disturbing effects of convection. In the North Island waves are created by the Kamai range, which although only 2000 ft in height, will carry gliders to over 30,000 ft. The range runs 15°W of true north and will produce waves in both SW and NE winds. A typical SW'ly situation is shown on the chart (Fig. 11.1). In SW winds the upflow can be reached either from an aerotow or by hill-soaring to as great a height as possible on the ridge, and then crossing over the top to the lee side. If, however, the upgoing part of the wave is not located by this manoeuvre, the glider will be dumped unceremoniously on to the ground in the turbulent flow to the lee of the range – fortunately there are some good fields.

NE winds give a stronger but shallower wave, the strength being due to the steeper fall away of the lee (west) side. The lesser depth results

—continued

from the wind shift at about 8000 ft which is usual in the easterly winds that precede the frontal cloud of a depression. It was in such conditions that a DC3 was lost with all hands while unknowingly penetrating the wave in complete overcast and rain; caught in the torrent of descending air it hit the rocks before it could clear the range.

In the Southern Alps of South Island, which rise in a huge barrier to the wind with 17 peaks above 10,000 ft, there is lift to over 30,000 ft in a westerly wind. Usually the 5th or 6th wave downwind is contacted from aerotow in the valley, and the system then worked upwind. The first wave is almost atop Mt Cook, 12,349 ft. The great heights which have been gained in this wave have been turned into distance, notably by S. H. Georgeson who used waves further along the range to beat the world Out-and-Return record with a flight of 625 miles. Keeping first cool and then warm is perhaps the biggest problem in high wave soaring, since the ground temperature at the start may be 40°C whereas for much of the flight −30°C is not uncommon.

Perhaps the most famous wave system in the world occurs at Bishop in the Sierra Nevada mountains, where a glider has reached 46,000 ft.

of rising air above and to windward of the crest of the ridge (Fig. 11.2). On the lee side the air moves on either more or less horizontally or it descends, depending on the shape of the ground. Generally this is all that happens, but in certain circumstances the air flowing over mountainous terrain will behave in a less simple fashion, creating huge waves which travel through the atmosphere, somewhat similar to those that occur in water. These waves in the air can be immensely powerful, sometimes exceeding 1500 fpm both up and *down*, and extending to very great heights. They have carried gliders to over 40,000 ft, but equally have pulled big jets into mountains. If we understand the behaviour of waves we can take all the benefits of their smooth upflows to save on fuel; and avoid the rough and insidious downdraughts. But to learn enough about them demands caution and a lot of thought as to how the air is behaving. In a glider we can have a marvellous ride to great heights, but it does not need much of a judgement error to get it all very wrong.

For waves to develop the hill or mountain needs to have a big slope facing the wind, and a considerable fall-away on its lee slopes; also the air itself has to be relatively stable (Fig. 11.3). This is necessary in order to give the air a laminar quality, so that it will smoothly follow the general flow lines. Strong thermals, for example, would disrupt the flow, particularly if they were rising from the

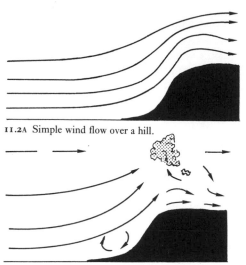

11.2A Simple wind flow over a hill.

11.2B Smooth wind flow is broken by convection upcurrents.

11.3A Typical strong wind flow over a steep ridge and (11.3B) over a more gentle hill.

windward slope which was instigating the wave. The most likely occasions for waves to develop are in winter, or when there is a temperature inversion at a height which helps to maintain, or even speed up, the smooth flow over the mountain.

If the shape of the lee slopes and the flow are suitable, the air coming over the top will pour down the slope into the valley and set up a rebounding motion, or wave. This may be damped out after a single oscillation or the motion may continue downwind of the mountain for perhaps 4–5 further rebounds. Usually the wind needs to be blowing at 20 mph or more for a wave system to become established; if it is light, and particularly if it slackens as height increases, the airflow, having descended the lee face will settle, and there will be no waves. It is the strength of the wind as well as the shape of the mountain and the character of the airstream that determines the wavelength and amplitude of the system (Fig. 11.4).

The gentler and smoother the slopes of the mountain the greater will be the wavelength, and the smoother the air; often the main

11.4 The strength of the wind alters the wave length, and will change the location of upcurrents and down draughts.

upflow and downflow, away from the ground, is remarkably smooth. If the mountain is steep-sided, waves will occur more sharply, with stronger upcurrents and downcurrents, and increasing turbulence close to the lee face of the mountain and to the ground. The wind will be likely to strengthen during its descent down the lee slopes due to adiabatic warming, and this extra velocity aids the development of further oscillations within the flow. If the wind is very strong, but the depth of the smooth flow layer is shallow – perhaps due to the lid effect of an inversion – the wave will be turbulent; the flow becomes complex and may break down from time to time. Another situation that will upset waves is when there is a second hill parallel to the main ridge; at a certain wind strength the wavelength will be in phase with the two hills, and the waves will develop perfectly. Then the wind slackens or increases and the hills will cease to be in phase and the system collapses, usually into areas of confused or dead air. There will, of course, be a change in, or breakdown of the system when the wind swings to a different direction.

The height to which waves will go may be enormously greater than the size of the hill, particularly when they develop over relatively small ridges. However, waves produced by some large mountains may not penetrate down into the lower levels of the

11.5 When there is a continuous cloud sheet the down flow of air to the lee of the hill may provide a break.

If the shape of the hill is suitable and wind is strong a wave system may occur, giving lenticular clouds.

A wave system may also give a cap cloud over the mountain and turbulent roll cloud to the lee, as well as lenticulars.

Typical banner cloud to the lee of mountain.

valley, so it is not possible to reach the wave to soar in it unless the glider is towed up to perhaps 6000–10,000 ft to start with. But this does not mean that flying in the lower levels will be free of turbulence or downdraughts.

If the rising air in the upflow is cooled sufficiently clouds will develop in and a little below the crest of each wave. These will be the almond or lens-shaped lenticulars that were illustrated on page 85. Underneath the waves, near the ground, the air is likely to be extremely rough; under the crests the flow often rotates completely with consequent sharp wind changes and violent turbulence. If cloud develops in the rotor the circulation can be clearly seen,

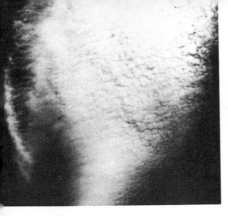

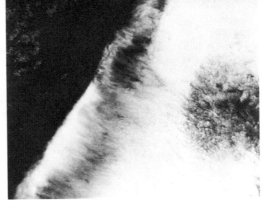

A

B

C

D

11.6 Lenticular clouds

(A and B) Looking upwards at a
lenticular. The windward edge is
on the left, and the difference in
texture between this and the lee,
downflow edge is clear. The
cloud was located 1½ miles W of
Cross Fell in the Pennines.
21 June. Wind E, strong.
1400 and 1445 hrs.

(C and D) Decay of a large
lenticular through which the
wind was blowing from left to
right. The cause was a slight
slackening of the wind, although
not enough to affect the new
small lenticulars at the bottom of
the photograph.

continued

As cumulus die away in the early evening there is more chance of waves becoming established. The photographs show lenticulars in a wave system which sometimes develops from Dartmoor in a WNW wind. (E) (1805 hrs) shows the first wave cloud appearing, (F) and (G) (1830 and 1910 hrs) show the stabilising air permitting the growth of perfectly shaped lenticular clouds.

E

F

G

and it looks alarming. This cloud is aptly named the roll cloud, and it has had more than one aeroplane on its back, including a glider being towed. The roll cloud may be substantial in appearance, like a long sausage, or merely consist of swirls or rags of clouds which rapidly form and dissipate. If the only way to reach the upflow of the wave in a glider is to tow it from an airfield downwind of the mountain, there may be difficulty in getting past the rotors to the upflow, particularly if the air is dry, so that the rotors are cloudless and cannot be seen. Attempting to tow through underneath them is least safe since the turbulent influence may extend down to the ground; it is better to go over the top, although much more margin than might seem necessary should be allowed.

The third variety of cloud typically associated with waves is the so-called cap cloud. This often appears on the summit of the mountain causing the wave, but it can, of course, also form when no wave is present because it is orographic cloud; it forms in the cooled air which is lifted over the ridge and evaporates as the air descends again, although the cloud may droop in the wind a little way down the lee slope. The cap cloud may also be called the Föhn Wall (page 225).

If the air is very moist cloud may form in the wave system so extensively that it is difficult to locate the wave clouds in the general mess; sometimes the whole area is solid with cloud except perhaps for slots of clear air or thinner cloud corresponding to the down-going part of each wave, especially the first wave, which receives the full benefit of the adiabatic warming down the lee slope. Since in these conditions it may be almost impossible to tell where either the mountains or the downdraughts are, we will be better off somewhere else, whether in a glider or aeroplane. In really big mountain areas even 10 miles away downwind should be regarded as suspect territory when the wind is strong, even if no obvious wave clouds can be discerned.

If the air is dry the sky may remain completely clear of cloud although powerful waves are present. More usually, however, the forced ascent of air in the wave is enough to wring out any moisture it contains, so at least some lenticular clouds are likely. They will remain in the same place over the ground because the wave system which created them is anchored by the mountain. At most the lenticulars will creep periodically downwind, stretching the wave length, and then quite suddenly jump upwind again to their original

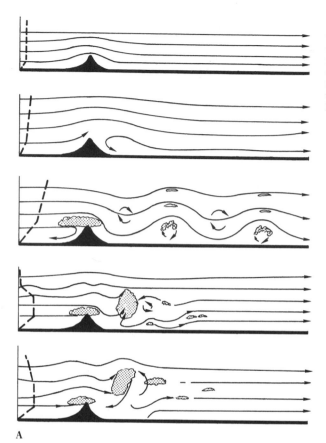

11.7A The change of temperature with height will affect the shape of the wave flow and the cloud produced. The heavy dotted line at left, indicating temperature, moves to the right where the temperature increases with height (inversion).

A

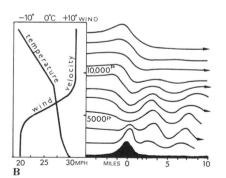

B

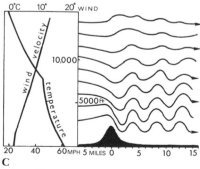

C

Now we build in both temperature and wind, (B) shows the existence of a stable layer between 3000–8000 ft in which there is a marked increase of wind, as might be expected in Föhn conditions. Above this height the air becomes less stable and the wind remains constant in strength. (C) shows a more continuous reduction of temperature with height, with the wind increase both steadier and stronger, and resulting in a more regular wave pattern.

A

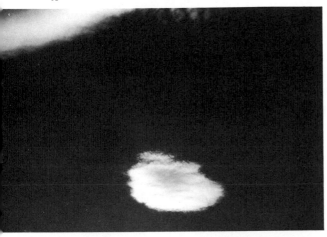

11.8 Lenticular clouds

These patches of lenticular cloud appeared and disappeared with remarkable rapidity.

The waves caused by the mountain were strong but, due probably to the steepness of the face, and/or fluctuations in the wind, the wave was erratic.

It would be difficult to work out where the wave was, and the risk of flying into the powerful and turbulent downflow would be considerable.

19 June. 2000 hrs. Wind E, strong.

B

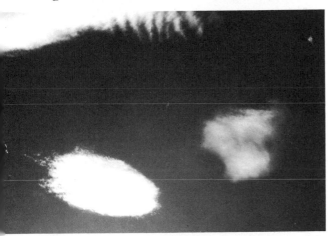

C

positions. This can be embarrassing to a glider pilot who is soaring in clear rising air on the upwind side of a lenticular when it jumps him from behind; without warning he is enveloped in cloud with some risk of icing if he stays there long. The apparently obvious way to get out is to fly further upwind, which would bring the aircraft out into the upcurrent, whereas downwind would take it into the downcurrent. However, it may not be so easy to go upwind. Because the wave is stationary relative to the ground, the aircraft will have to be flown at a speed in excess of that of the wind at that level to alter its situation relative to the wave. At, say, 20,000 ft the wind could easily be blowing at over 100 knots, and some aircraft may simply not be able to fly fast enough. If this is the case, the cloud should be left by flying downwind; trying to get out across wind will simply extend the time spent in cloud.

If flying along the valley or trough between two wave clouds and wishing to regain the upcurrent as soon as possible, it may be better to skip half a wave length downwind, rather than struggle into the wind.

Soaring in a powerful and smooth wave can be very satisfying, and has occasionally been done in quite heavy aircraft, such as the old P38 Lightning which weighed some 8 tons. It is sometimes so enjoyable that a hazard far removed from the wave can be overlooked – the weather near the ground perhaps 15,000–20,000 ft below. Particularly late in the day we should watch for signs that cloud or ground mist might obliterate our chances of a trouble-free landing. Cloud may develop simply as a result of a slackening of the wind and cooling air, or because moister air is moving in, perhaps from the sea. If the sky is clear so that there is a chance of considerable end-of-the-day radiation, ground mist should be expected. Once the first wisps have appeared of either low cloud or mist, there is a real chance that it may grow spontaneously everywhere at once; if conditions are right complete cover over a large area can take place in less than 10 minutes. Another hazard for the forgetful enthusiast is that it will become dark on the ground while the aircraft is still high in sunshine. It takes a surprisingly long time to descend in a glider from, say, 15,000 ft, even with airbrakes out. It takes even longer in some aeroplanes and darkness does not wait.

Since waves make it comparatively easy to reach great heights we should pause for a moment to consider the matter of oxygen. Although a fit pilot can remain reasonably sensible for a minute or

11.9 Lenticular clouds

A A sky full of lenticulars caused by waves to the lee of Cross Fell in the Pennines. The extensiveness of the system could have been due to some of the Lake District mountains also being in phase with the wave.

Since the weather is otherwise good and the clouds high above the mountains, there should be no real problem flying through the area. Should height be steadily lost in a down current, the aircraft should be flown into wind or downwind (across the bars) until clear of it.
June. 1200. Wind E, very strong.

B Waves over the Lake District in a W wind (picture looking south). The system is weak and there are a lot of lower level cumulus.
22 July. 1700 hrs.

C A gentle wave system with the wind from left to right. Below is cumulus and strato-cumulus which earlier had been more extensive and given an undulating appearance by the wave.

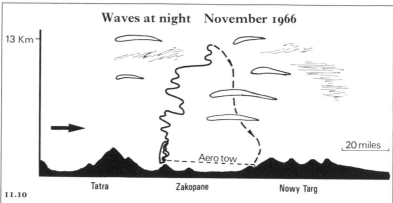

Waves at night   November 1966

13 Km

20 miles

Aero tow

11.10   Tatra   Zakopane   Nowy Targ

That lenticular clouds can be recognised at night is shown in this account by Josefczak and Tarczon, two Poles out to beat a record. They had a Bocian glider, fur clothing and oxygen, but no cabin heating.

We took off from the airfield at Nowy Targ hoping to break the two-seat height record by flying into the night. This required a rather special approach because of the different look of lenticulars in darkness. During the afternoon the 7/8 strato-cumulus started to disintegrate and lenticulars appeared at a considerable altitude. At take-off only one rotor was left east of Zakopane, giving lift from 200 m above airfield level up to 2300 m. Take-off was made in a fairly strong wind at 1524 hrs, and the tow made at a low height in turbulence towards Zakopane. . . . There I released at 1700 m above airfield level in lift of 8 m/sec., climbed to 3500 m, and then on the side of almost the last rotor descended as far as possible to get a low point on the barograph from which to start the real climb. I managed to get within 150 m of the terrain, and then started to climb in the lift. At 4500 m I left the rotor and moved into the wave proper. When it started to weaken I flew towards Zakopane. Whilst doing this I was climbing at 4 m/sec. with the wind from 190 deg. I managed to reach 11,500 m in an area of lift 8 km long. When this lift started to deteriorate I flew again towards Zakopane and reached 12,500 m in lift at 3 m/sec. It was now dark. All the lower cloud had disappeared and only lenticulars remained looking like black strips in the starlit sky. I started my descent at 1800 hrs and landed at 1840 hrs in calm air.

so at 20,000 ft, it is generally accepted that flying above about 14,000 ft without oxygen is unnecessarily hazardous. Shortage of oxygen even for a brief period affects us both by impairing our reasoning ability, and by worsening our eyesight and resistance to cold. The first is more important because anoxia allows decisions to be made without our realising how stupid they are; for these reasons it is better to use oxygen above about 10,000 ft. After all, at this height a breath of a certain capacity will only take in three-quarters of the normal quantity of oxygen, and at 20,000 ft only half. Aeroplanes are not often soared in wave lift, but may be gradually forced to go higher and higher in an attempt to get over the top of thundery turbulent cloud. Without oxygen we should regard 15,000 ft or over as just as great a hazard as the cloud itself,

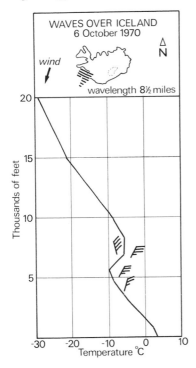

11.11 Waves in Iceland

6 October 1970
Powerful waves develop all over SW Iceland in a North to NNE wind. The lenticular clouds formed are prominent enough to show on satellite photographs. To reach this wave in a glider it is launched into hill lift near the club site at Sandskeij, and by climbing as high as possible in the slope lift the waves can be reached, and may then carry the glider to 20,000 ft.

Both the main airports are within the wave influence, so jet pilots have come to know it well.

The temperature at higher levels shows a strong inversion from 5500 to 8000 ft which would assist a laminar flow of air over the mountains upwind.

and abandon any continuation into either of these situations.

Waves are least likely to develop on summer days of strong convection because of the need for the smooth laminar airflow in which they can become established. This does not mean that they will not occur when there is any instability, but that the flow must be reasonably smooth for the wave oscillations to develop fully. Thermals in the surface levels under, for example, an anticyclonic inversion, will not necessarily prohibit waves, although they may cause them to be somewhat intermittent or have only a short life. When thermals are strong but conditions otherwise right for the appearance of waves, they will establish themselves late in the day as the thermals weaken and die away, probably continuing until dark or later. A good time for waves to develop is in the smooth air of early morning, but they will often collapse as convection gets going.

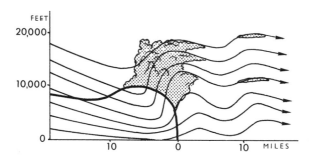

11.12 Waves can occur ahead of a cold front.

We sometimes see wave clouds in the sky which could not have been caused by mountains, because there are none in anything like the right place. These can be produced by a sharp increase of wind speed at some height within the general airstream, in rather the same way as a tide causes ripples or waves in underlying sand. The sort of weather situation in which such waves occur is ahead of a depression; well in advance of the warm front where there are appreciable changes of wind and temperature in the medium and higher levels. They may also occur a shorter distance ahead of a cold front (Fig. 11.12). Such waves are, of course, not stationary relative to the ground but drift along with the air mass in which they have been formed. The waves themselves may be shallow, or they may develop through a considerable depth with lenticular bars lying across the flow. The deeper waves which sometimes appear in the medium levels, around 10,000 ft or so, may have an amplitude sufficient to cause downwash gusts to reach the ground. These gusts may be powerful and arrive without warning, and it is easy to get caught out by them because the mean surface wind on such a day may be neither noticeably strong nor rough. Each year in Britain one or two gliders and light aeroplanes are damaged by wave gusts, because they have been left in circumstances which seemed quite safe. When drifting wave clouds can be seen in the sky it can save a lot of trouble if any aircraft which are to be parked unattended are secured. Over Britain moving waves seem to develop most often in W and NW winds.

Clear Air Turbulence, or CAT, that continuous or intermittent bumbling roughness that one gets riding in a passenger jet up in the sunshine, is similarly caused by ripples and waves over the shear face between air of different temperatures and moving at different speeds.

Part 4　Flying out of trouble

12.1 Squalls

A

A A small cu-nb growing in moist unstable air. The wind is strong and increasing with height, shown by the 'lean' or slope of the cloud. Heavy rain is falling from the low base.

B Cloud associated with a cold front among hills. Almost as soon as the moist air is forced up over the hills more cloud forms. The strong wind causes it to be disorganised with a ragged base.

C The lower, dark, cloud is forming orographically below the hilltop from which the photograph was taken, and rising to mingle with the low ragged frontal cloud blowing across the top.
July. 1130 hrs. Wind W, 20 knots, gusty.

B

C

12   Weather mixed up with the ground

Unlike a jet captain who spends much of his time above the weather complete with all the aids, in a light aeroplane we are mixed up with it all the time, and sometimes the weather is mixed up with the ground. Although many small aeroplanes are equipped with navaids these can never be the whole answer to safe flying in bad weather; apart from the fact that it will always be possible for the best equipment to fail, it is still the pilot who has to take the decisions, and do the flying. To operate an aircraft safely in extensive, turbulent cloud, with poor weather underneath, or at night, demands a high degree of skill and continuous practice. The possession of a current instrument or IMC rating is no guarantee that we will be able to cope effectively in the first bad winter conditions, after a long hot summer with no instrument practice. A pilot with only minimal instrument flying experience, or even without any at all may not, in fact, be necessarily worse off, provided that he knows his limitations. For the many of us who fly small, or vintage, aircraft, and who may not have the opportunity to get airborne as often as we would like, knowing how to stay safe in poor weather becomes an essential skill. Sooner or later there will come a moment when the correct weather interpretation means survival, and a mistake is too expensive. The essential need is to avoid the irretrievable situation; in other words, to turn back, divert, or land before it is too late. Obviously an experienced pilot will, or should, be able to deal with worse weather than a student pilot, not only because he will have developed some cunning over the years but because he will be better at sorting out an emergency landing into a strange field. In deteriorating weather, as soon as cloud base has descended to 1000 ft above the ground, the inexperienced pilot is at risk, because there is a good chance that he will mess up a field landing if he has less than this height in which to do it. An instructor is not likely to send off an early cross-country attempt in poor conditions, but should an inexperienced pilot find himself unexpectedly in worsening weather, he should swallow his pride, turn back or, if necessary, land at the nearest airfield while he still has at least 1500 ft above ground.

Weather problems met with in flight nearly always arise from a

Narsarssuaq

After several days delay at Goose Bay it now appeared that the weather was go for the crossing to Greenland. The progs were showing improvement and indicated excellent weather on arrival.

An hour after departure I made contact with an incoming pilot who informed me that the weather on his departure from Narsarssuaq was improving and should be as forecast on my arrival. He was returning from holiday in Europe with his family.

I was ferrying a pressurized Navajo at Flight Level 170 with a TAS of 196 knots and groundspeed 230 knots. My ETA Narsarssuaq was 2026Z with sunset at 2050Z. You have to be on the ground there by official sunset and if you ever saw this place you wouldn't, for one moment, question this policy. I reviewed the forecast, and it looked like a sunny arrival.

Contact was made with Narsarssuaq with a request for descent clearance, present weather, and the altimeter setting. I'll dispense with the description of shock. 'Weather 400 broken, 2000 overcast, 2 miles, light rain, light snow, wind 210/16, temperature plus 3, altimeter 1002 mb. Fjord closed, fog. What are your intentions?'

With no available alternate airfields, this left two choices. I could put this bird down on the ice cap visible off to my left or make a hair-raising let-down into the fjord, which would be nearly dark at field level with the present weather. I elected my original landing place – the fjord – as the more alternatives I dream up the more confused I become.

I had previously been cleared to FL70 and was approaching this altitude when I observed a break in the cloud not 5 miles ahead. The water was visible but not too clearly. A request was made for a visual descent and approved at the pilot's discretion. I reduced power, lowered the landing gear and 30° of flap. I estimated that I had approximately 15 miles to the airport. The weather was as indicated, it was getting very dark and visibility was at or less than 2 miles. However, not only had I let down into a sucker hole, I could see by the icebergs that I had let down into the wrong fjord. 7000 ft is the recommended altitude in this area, and an emergency course reversal and climb out soon got me back there. Narsarssuaq was frantically trying to call me, but the lack of contact was due to the 2000 ft high terrain between the fjord I was in and the fjord with the airport. I was now totally committed; the time was 2035Z, with one hour and fifteen minutes fuel remaining.

It was evident that the operator at Narsarssuaq was an old pro. From the tone of his voice he knew what I was up against. He asked me to report over the radio beacon at FL70 and immediately turn to a heading of 304°, and he asked for my maximum rate of descent that could be safely maintained. I replied 3000 fpm. Everything was set up with IAS

—continued

120 knots – but for what?

'Start your descent . . . pause . . . now. Report through 5000 on ONH 1002 mb. Now turn left to heading 240, and report through 3000.' I acknowledged and switched on windshield heat. Light rime ice was forming and I needed all the visibility I could get if, and when, I broke out of this stuff.

'Turn further left, heading 200.' Then 'Turn further left heading 160, report through 1000 ft and when visual.'

Even with the old pro and all his electronics down there guiding me, I felt mighty weak when I reported through 1000 ft and no ground contact. I broke out of all cloud at 400 ft on base leg for runway 08, a half mile from touchdown. We were on the rollout at 2043Z, 17 minutes late.

I couldn't wait to thank the man who brought me down to a safe landing, but he would have to wait, as I sat there for nearly ten minutes before I could get my legs to comply with my wishes. The rear door was opened and the light shone on the face of a young man who introduced himself as Per Andersen and invited me to ride along to the operations room. He was filling in for the regular man who was on holiday. Then I got the second shock of the day. There was no DF equipment or radar. Per had taken the microphone outside so that he could hear the aircraft overhead and kept the sound of it within hearing range. If the pilot kept him informed of his altitude, what more could be done! What more can I say but thanks to Per Andersen, private pilot, and one damned good controller.

—From an account by Eugene E. Locke, *AOPA Pilot*

mixture of reasons, including failure to appreciate that the weather is deteriorating, and inadequate flight planning. The first action, therefore, is to obtain the fullest and latest met information before taking off on any flight for which the weather is not known, or seen, to be acceptable all the way. We know that this does not mean simply receiving a typed piece of paper or being shown a synoptic chart, but studying both, and continuing to ask questions until the situation is fully understood (check list, page 116). It is necessary to be able to visualise the weather in the mind's eye, because three things can subsequently go wrong; aspects of the forecast may be incorrect (not often, but it can happen), local differences from the general weather may be encountered, and – for a variety of reasons including getting lost – the route may change. So accurate recognition of what the weather is doing is essential; because it will permit sound flight decisions to be taken in plenty of time, and not when safe alter-

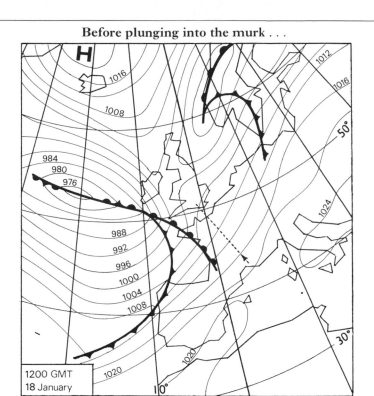

Before plunging into the murk . . .

1200 GMT
18 January

12.2   18 January.

Most pilots obtain route forecasts before starting a long flight, but sometimes the description of expected conditions is no substitute for a personal, and thorough, study of the synoptic chart.

A pilot, flying from the south of France to London in midwinter obtained a forecast in the sunny south which gave reasonable weather for the route, including the terminal region, but said that a warm front lying N–S across Britain might affect the situation. He flew over the Channel under fairly low cloud and poor visibility, finding conditions deteriorating more rapidly after crossing the coast – particularly over the South Downs. He was in contact with Control at his destination, but concluded that the weather was now so bad that he should return to the coastal low ground. Unfortunately, in doing so, he hit the north face of the Downs, about 650 ft ASL, and was killed.

The actual weather at the time was 4/8–7/8 cloud varying from 400 to 700 ft, tops 800 to 1000 ft. Above this was 8/8 cloud in layers from 1200 to 6000 ft. Visibility below cloud was 3–5 km, reducing to half this in rain. There was also some hill fog. Surface wind was SW 15–20 knots at sea, 10 knots inland.

—continued

The South Downs lie 3–5 miles inland along an approximate E–W line, broken by valleys and spurs. They rise to 800 ft ASL. In such conditions, if a pilot studies the synoptic chart, along with his flying topographical map, what information can he supply himself with?

1. The chart on this day shows a deep depression centred WNW off the British Isles. It is partially occluded, but the warm front is still ahead of the cold front over SW England. The warm front is moving in a direction which WILL cut off the destination, although the timing is not known. A calculation of the speed of the front would indicate: 150 miles to go at, say, 20 knots at worst=about 5 hours. From the time of the chart (12 GMT) there is, therefore, a real risk that the arrival of the aeroplane and the arrival of the front will coincide.

2. An active warm front WILL influence the weather right down to the ground. Therefore even quite low hills will (a) be mixed up with it and (b) almost certainly result in very low orographic cloud forming on the windward side. There WILL be turbulence and downcurrents on the lee side, the severity of which will depend on the strength of the wind.

Answer so far. Safe arrival, flying contact below cloud, is only possible if the aeroplane beats the front. This is uncertain, so:

(A) A continued visual appraisal must be made of the weather to ascertain the location and progress of the front, and

(B) A practicable diversionary plan must be made.

If on arriving over the Channel the weather is found to be deteriorating, it MUST be assumed that the effect of the land will make matters worse. The only alternatives are, therefore, to land at an airfield on the coast itself, or to fly back, south-eastwards, into France and land there ahead of the front.

natives are being shed faster than we can think. Investigation of weather accidents has shown that few are due to the forecast being wrong; most result from too superficial an interpretation of the met information, or from dithering about what to do. Statistically, daytime weather accidents occur with the greatest frequency in a warm, or occluded, frontal situation; yet visually such weather gives the greatest warning of its approach. So what goes wrong?

The pattern of such accidents indicates that the pilot either continues to fly under lowering cloud until it almost reaches the ground, gets lost in the mist and drizzle, and finally hits something; or he decides to go over the top, and flies into high ground during the descent.

Let us study this weather problem in more detail through the eyes of a pilot flying visual – our eyes, on a route towards an approaching Low and its warm front; which can be seen to exist by the cirrus already overhead. We are not worried since there are diversion airfields on the way and we have wisely ensured that we are in no hurry; but the map shows that there is some high ground towards the end of the trip – up to about 1500 ft. We take off and fly at 2500 ft AGL for a couple of hours. Without radio we cannot ask for an altimeter pressure correction, but know that the altimeter will slightly overread as the air pressure lowers towards the Low centre. We make a mental note to watch height on the approach to land so as to avoid making too low a circuit in error. On we fly under imperceptibly lowering cloud and reducing visibility, and after a while decide to descend some 800 ft as this seems to make map reading easier. There is little wind, the air is smooth, and the windscreen clean. On approaching the rising ground ahead we decide to take a bit more height, but discover that in the meantime cloud has come down to our level; so we peer ahead, see no big problems and continue over the first lot of hills. The next ridges are higher, although they look as though they will remain clear. But over the windward face of the first hills cloud unexpectedly lowers – due to the orographic effect – which reduces our room for manoeuvre. But we know where we are – having folded the map sensibly before starting – and the tops of the next hills are still visible. We pass over them with about 100 ft to spare just as the drizzle starts to blur the windscreen. This drizzle masks the lowering cloud and annoyingly we fly into it while doing a quick map check. The quick dive out does not use much height and there is only one more lot of high ground to cross, with a through valley almost on track. In the poor light and increasing rain it is difficult to see, so we let down a bit more in order not to miss the valley, but are now below the tops of the hills on either side, and they have become invisible in the murk – or cloud? On some of the nearer slopes dirty little orographic rags are forming, a quick look round shows that the weather has still not closed in behind; but a pylon looming up in front a bit too close for comfort, is scaring; the urge to hurry on and clear the hills becomes irresistible. Suddenly everything vanishes; ramming the nose down to clear cloud, trees rush past underneath. Then the ground starts to fall away and we are through. We land in steady rain at our destination not sure whether to tell all, or say nothing.

We were not skilled, but lucky, simply because on this occasion the wind was light. If it had been stronger orographic cloud would have developed more rapidly and extensively, turbulence would have been considerable, and the reduced ground speed (we had a head wind, remember) would have meant that the aircraft was longer among the hills than previously in the deteriorating weather. In these conditions a last ditch attempt to turn back is too often frustrated by cloud down on the deck, and across the retreat. Flying in really poor weather is sometimes acceptable if the weather stays still. It is the strong wind, rapidly shifting, situation in which low cloud grows and dissipates, rain starts and stops, and visibility improves and becomes nil, that is the killer. Planning to fly contact along a cloud covered valley in such conditions is as chancy as winning a big lottery. If in any doubt, rough air and the appearance of wisps of orographic cloud quite clearly spell 'Go back now'.

Weather accidents are not always a matter of just flying into a cloud smothered hill, more often the aircraft has been flying low half in and half out of broken swirling cloud for some time without any clear decision being taken whether to stay visual or go on instruments. Since the cloud base is irregular, giving little indication of the line of the horizon, which in any case probably keeps on disappearing, there is a real risk of becoming confused or disorientated. We can be convinced that we are flying straight on the course which will take us out over low ground and between hills, and be looking out for signs of the valley, whereas we are gently turning, and suddenly cloud with its hidden hills is all around.

A bit rough . . .

17 MAY 1939

The memory of this flight still gives me nightmares. Five of us had flown passengers from Weston-super-Mare to Cardiff to watch a big fight, and shortly we were to fly them back again. At about 9 pm the weather started to deteriorate very rapidly and as one of the other pilots remarked after a quick inspection, 'It's a bit rough.' This was a masterpiece of understatement as the cloud base was about 300 ft or less, it was pouring with rain and the visibility couldn't have been more than half a mile, if that.

—continued

The Chief Pilot was far more experienced than the rest of us, so he was first to take off and very quickly vanished into cloud. We had all decided our plan of navigation, which was to climb to about 800 ft, turn on to a heading of 200 degrees, holding this for between 7 and 8 minutes, and when receiving a bearing from Bristol of 050 degrees, turn on to this heading, letting down to become VMC over Weston pier. This would keep us clear of Worlebury Hill (400 ft) about two miles to the west of the field.

On the ground in a comfortable lounge this sounds very professional and comparatively simple, but when one has the vast experience of 5 hours under the hood in a Tiger Moth on basic instruments it becomes without any shadow of doubt one of the most hazardous operations known to aviation.

I was flying, if that is the operative word, a DH Dragon equipped with only basic instruments and one of the old-type altimeters which were not accurate below 200 ft, and which presented extreme difficulty in reading in the prevailing conditions. Anyway, off I went following the others in strict seniority, complete with eight rather inebriated passengers all of whom were fortunately relaxing into a deep coma.

My instructor once told me that instrument flying was simple if you held the stick between forefinger and thumb and rested your elbows on your knees. I would have had news for him that evening. However, having managed to climb to 800 ft I tried to contact Bristol. As we always had extreme difficulty in raising them under the best conditions, it was no surprise to find that there was a deathly hush apart from my own shrieking into the microphone. I abandoned them and decided to stick to dead reckoning.

Suddenly to my astonishment and alarm I noticed the aircraft was rapidly descending in very steep turns, the compass giving an excellent demonstration of an anemometer in a gale. After considerable concentration I managed to level the wings and promptly soared up to 1000 ft. A few seconds later the aircraft decided on a repeat performance even better and steeper, and this it did several times. Opening the sliding window in desperation I peered into the gloom trying to find some form of pinpoint whilst being blown about and completely drenched by the rain lashing in. While I was wondering what on earth to do next I happened to see faintly through the clouds some lights. Fortunately they were on Weston pier and it became desperately necessary to become VMC for fear of this priceless piece of information being lost.

The next manoeuvre would have frightened the living daylights out of sober passengers, apart from the residents of Weston, who must have been startled to see a rather ancient aeroplane come hurtling out of the clouds and flash across the pier at 150 ft at a speed far in excess of that permitted by the makers. No one was more surprised than myself to find

—continued

If the front is occluded the worsening weather will, of course, possess the nastiness of both warm and cold fronts. The turbulent thick cloud and heavy rain of the latter will be masked by the warm front drizzle preceding it; and the increasing cloud thickness will cause the light to worsen. If the aircraft, hills, and an occluded front coincide unexpectedly, there may be remarkably little time to escape. The thick cloud will of course also curtail the daylight and latest landing time by half an hour, or even more.

Squalls

Weather brought by the cold front – whether it be a fierce line squall bludgeoning its way across the country or just a tiresome spell of cloud and rain – has been described earlier. To some people the phrase 'after the front has gone through' means that the bad weather has finished, although when the air that follows is cool, moist, and windy, the meaning is mostly fallacious. This is the sort of thing that happens. We decide to delay leaving until the forecast front has cleared past, which it does exactly on time. One up to the met men, we say, and take off into a fresh wind, brilliant blue sky, with visibility unlimited. Twenty minutes later some attractive and sunlit cloud appears along the horizon, which with surprising speed clambers up the sky to reveal a dark and stormy barrier. It no longer looks pleasant, and it is not; the sensible pilot will quickly decide that its turbulence and heavy precipitation are not for him. If there is not an obvious way around the end of the cloud, or through a big enough gap (a small one may close up as we arrive), we should hurriedly return to somewhere that will give us enough time to picket, or hangar, the aircraft before the first gusts of the storm arrive. Such squalls, developing in unstable air, can be violent, particularly in the spring and early summer. They are however less common in autumn when the land, sea and air are all more similar in temperature. Since they are caused by convection they will lessen or die out in the late afternoon or early evening, leaving a clear night. Unexpected squalls or big 'April showers' can be difficult for a student pilot, far from home on a navigation exercise. It is easy to be misled by the clearness of most of the sky, and become tangled up with a squall while trying to be super-obedient to the designated course.

Quite frequently squalls come in waves or lines. They are often narrow, but the rain or hail they throw out may be dense enough to mask both the lowering cloud base, and ground features – like television masts. Since such storms often travel a little faster than the mean wind speed, as well as growing rapidly at the same time, it is usually better to divert around the upwind end, rather than be pushed further and further downwind. It can be taken as one of the meteorological Murphy's Laws that the more sharp the clearance following a cold front in springtime, the more likely is it to be followed by further waves of cloud and rain, until the evening. If the rain just stops gently under a grey sky, it is improbable that further squalls will occur.

In maritime countries like Britain and the Netherlands squall lines of great length and severity are not so common as over large continental land masses, such as USA and central Europe. If there is an active and extensive squall line cloud barring our way, it is impractical to attempt to take a small aeroplane over the top, unless it, and we, are equipped to go to 25,000–30,000 ft. Since the ceiling of the aircraft will almost certainly be less than this, we may be tempted to push through the cloud itself to the other side. The usual result of such a manoeuvre is that we will be forced to stay on instruments for a length of time which proves a sore trial to our powers of concentration. If the extent of the cloud is not known with certainty – and it usually is not – the best chance of getting out of it soonest will be by flying away from the squall line, preferably downwind, and descending carefully.

Snow

In winter, if the route is more or less parallel to a warm front, or under continuously precipitating cloud, rain may change to snow, or vice versa. This is likely to happen where air is being orographically pushed up over higher ground, when the flying height is increased, or when flying towards colder air. There may be snow ahead of a warm front where the precipitation is falling through the cold air, but, as has been mentioned, nearer to the front itself the air will be warmer, and the snow often turns back to rain.

Hills covered by snow under complete overcast may be indistinguishable from the sky, until we suddenly see a house on a cloud. If the previous low ground has been free of snow, it is easy, initially, to miss seeing this snow-covered high ground.

Maps

It goes without saying that a suitable map, of not less than 1 : 500,000 scale, should be carried on every flight away from the circuit area. It should be a personal possession, and not a map kept in the pocket of the aircraft which may be discovered to be of Corsica, when what we want to know is the height of the Welsh mountains hidden in cloud all around us.

Icing

The first thing about icing is to be aware of when and where it is likely to occur; so it is useful to be able to make a quick approximate calculation of the height above the ground at which the temperature will drop to zero C in cloud. Such a calculation is simple, if we remember the adiabatic lapse rates. If, for example, the temperature on the ground is 12°C and cloud base is at about 3000 ft, the temperature at this height will be 3°C (the dry adiabatic lapse rate of 3°C per 1000 ft × 3). Thereafter the lapse rate will be at 1·5°C per 1000 ft. From the cloud base temperature of 3°C a further 2000 ft will give us freezing level (the saturated adiabatic rate of 1·5°C per 1000 ft × 2). So freezing level will be about 5000 ft above ground. In practice it will probably be a little higher than this because the clear air may not be cooling at quite the dry adiabatic rate, but as far as flying is concerned, any such error will be on the safe side.

Icing will occur when the aircraft is wet or in moist air and is cooled, even if there is no precipitation. In temperatures between 0°C and −10°C supercooled droplets will freeze on the aircraft – including the windscreen.

Carburettor icing

Present generation aircraft piston engines, whether possessing a carburettor or fuel injection, are susceptible to icing. Trouble is most likely with carburettor engines, but fuel injection engines are not completely immune. All such engines, other than those used on the most elementary types of light aeroplane or motor glider, are fitted with a pilot-operated hot-air intake control. This is normally sufficiently effective to enable the engine to run satisfactorily under icing conditions, and to clear a limited amount of ice. But it will not restore full power quickly if the engine is badly iced up.

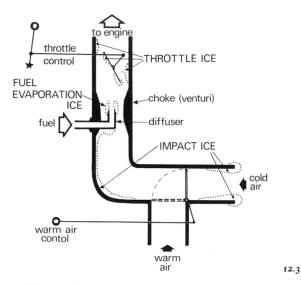

12.3

The engine cannot be run continuously on hot-air intake as this would reduce power output and increase fuel consumption. So we have to live with the problem, although it is hoped that before long manufacturers will have in production a simple and efficient way whereby the engine senses the onset of icing, and takes its own preventive measures.

Icing will occur under most circumstances when the temperature is moderately low and the humidity is high, and the ice will cause trouble in three different ways:

1. Impact ice, which may build up on the air intake or immediately inside it. This is unlikely to be serious initially, and in any case

we have warning of what is happening since we will see similar icing on the leading edge of the wings.

2. Ice may form inside the induction system as a result of the reduction in pressure which takes place when the engine is running on part throttle.

3. Ice may build up as a consequence of the actual evaporation of fuel into the air intake stream, lowering the temperature to such an extent that the moisture will condense and freeze. For example, the overall temperature drop in the carburettor may amount to 30°C, so that when flying in a clear humid atmosphere even as warm as 25°C ice may form even though there is no cloud present.

Icing often first appears when the engine is throttled back. It is also likely to occur on the ground during take off on cold damp days, or during descent in or near cloud. The symptoms are loss of power and rough running. Using the carburettor heating – and this should be either full on or full off unless an air intake thermometer is fitted – may appear to make matters worse; the melting of the ice inside the induction passages of the engine may result in it running even rougher. The temptation to return to cold air should be resisted, and the hot air be allowed time to clear the ice.

Actions: (fixed-pitch propeller aircraft)

1. Before take-off select hot air for a few seconds and then return to cold air intake, keeping the throttle steady. This should result in a drop in revs when the heat is engaged, and then a return to the original value. If erratic results are obtained it is probable that carburettor icing is occurring. Re-select hot for at least 30 seconds to clear the ice, then return to cold before take-off, checking that the correct static rpm is achieved.

2. While cruising check occasionally to see whether icing is taking place by engaging hot air. If icing is suspected engage hot and keep it there even if the running temporarily gets worse.

3. Do not make prolonged descents with the engine throttled right back, as the amount of heat available may be insufficient to prevent ice forming even with hot air selected.

4. Before landing remember to change to cold air in case full power is needed to go round again.

13.1

13   Not always lost above cloud

There are many occasions when the air is clear above or between layers of cloud, and flying there gives a pleasant and smooth ride, provided that the descent can be made without risk. It somehow happens, however, that we are there without guarantee of a safe let down: cloud may have been entered intentionally or inadvertently from below or it may have been entered from above because all the holes disappeared. When waking up to the fact that the aircraft is now over a dazzlingly beautiful, but seemingly endless, detergent surface of unknown thickness, there may be an inclination to hurry down. Without radio it is prudent to pause and try to assemble some information before plunging into the murk. Our aim is threefold. To avoid, on breaking cloud, an area with inadequate terrain clearance, controlled air space, or the ocean without sight of land.

In terms of the general weather, it is obviously sensible to turn back if flying towards an approaching front. We know that the air ahead is likely to contain quantities of strato-cumulus and stratus which join up with the front, so the cloud will thicken, and the weather worsen, in that direction for some time before any improved weather can be reached. If following a front which has already passed through, and catching it up, again the best course is to turn back, or at least hang about until it is certain that the thicker cloud will have moved on beyond the area in which we wish to descend. If the air is moist, which extensive cloud cover indicates that it must be, it should be reckoned that the base will be low, so if there is hilly country underneath it is wise to change course for an area where the terrain is lower and flatter. Obviously, we should keep a note of any course changes and the times at which they were made.

Let us now assume that we have been keeping a log, and can calculate where we think that we should be considering the uncertain factor of the wind which may be stronger, or different in direction from that expected. We have fuel and daylight, have avoided any complex frontal clamps and now want to descend through cloud safely and as close to the right place as possible. Even if we think we are in the right place by our reckoning, we should check the evidence of our eyes. So what can we see?

The appearance of the cloud sheet top surface in the lower levels

A

13.2 Decline of a cloud sheet

A 8/8 cover of stratus which is probably 1000 ft thick – or more. The top has a cumuliform appearance resulting from the effect of surface warming.

B The sheet is breaking as we go south and away from the low-pressure centre. It has also been getting thinner, and is now no longer a threat.

C Farther away from the Low centre, the cloud sheet no longer exists, in fact it may have been clearer here most of the day. The haze held by the cloud level layer is noticeable.

B

C

is affected by ground features, particularly if there is a wind. A sheet of low stratus, for example, will be thickened and pushed up where there are hills, and flattened and thinned where it flows down over lower land or the sea. It may break along the coast or even look like a cascade. With so much geographical information likely to be available, the first thing to do is to climb as high above the cloud sheet as practicable and have a good look round.

Over Britain there are few mountains high enough to stick up through the several thousand feet of cloud which typically, and frequently, occur, but the mass of the Welsh mountains, for instance, will transfer a duplicate of their mass, and to some extent even their shape, to the top surface of the cloud. The cloud tops will be higher, with the surface more broken and 'rocky' over them, than elsewhere; the layer cloud may even develop a sort of cumulus appearance above the crests of the invisible hills. If there is a strong wind, undulations, waves or ripples in the cloud surface will continue many miles downwind of the hills. Lenticular clouds may be present announcing the development of waves. They may appear singly or as a series lying in the direction of the surface wind. These smooth-edged clouds may be merged with the cloud layer, or they may be at some height above the sheet on which they throw a shadow. If it is vital to know about the high ground underneath, the lenticular clouds should be studied with care. If the shadow is stationary in relation to the drifting cloud sheet it shows the definite presence of a stationary wave caused by substantial mountains. These mountains will be lying generally across the wind some distance – perhaps 3–4 miles – upwind of the first lenticular cloud. Lenticulars also indicate that the wind at the surface is strong. Absence of lenticular clouds does not necessarily mean that there is no wave system; if the top surface of the cloud sheet contains big parallel undulations, there almost certainly is, although it may not be so strongly developed.

If there are lenticulars or undulations in the cloud top the most dangerous direction in which to descend through cloud is towards them into wind – the surface wind. Should the aircraft encounter large surges of lift or sink during any descent into wind the presence of waves caused by mountains ahead must be assumed. If, in the event, there are no waves, no harm has been done. Descending when flying downwind may be safer since the aircraft will be moving away from the cause of the waves. There could of course be further mountains in this direction, perhaps in phase with the wave, apart from the aircraft having to traverse the ups and downs of the rest of the system. Turning across wind keeps the aircraft longer in the same sort of air, either in the upflow of the wave, or in the down, so if thinking time is required it is obviously better to fly across wind, particularly if already in the upflowing part. In practice it may be difficult or impossible to remain in a particular part of the wave since the aircraft must be kept over the appropriate bit of ground – and this is invisible. The main thing is that the risks of trying to descend through cloud in wave are very real, but they will be reduced if we are working out all the time what bit of the system we think we must be in, and where its source lies, rather than blundering about in uncomprehending hope.

When a sheet of cloud starts to decay it becomes thinner, with an increasingly smooth surface. It will remain thicker longer to the windward of hills, although as the sheet subsides and thins there is more likelihood of the tops showing through. Holes start to appear in the sheet, which initially may be surprisingly isolated; early ones may have been caused by the extra warmth from big towns. Again, with plenty of height above the cloud sheet it is easier to locate holes. If we are skating about just over the cloud surface a hole will be passed and possibly lost again before anything can be done about going down through it. Although there is some chance when flying higher of a hole closing up again before the aircraft can be fitted into it, the ability to get the speed, and maybe some flap, down, check on the terrain underneath, note the time in the log, and circle down in a calm frame of mind is far better than screaming earthwards at Vne-plus into some unknown cauldron.

Sometimes discoloration of the top surface of cloud indicates the location of a town, refinery or cement works, the discoloration being, of course, downwind of the source – on a windy day surprisingly far downwind.

If there is plenty of fuel aboard and land underneath, the chance of finding a break in the cloud will be increased by flying away from the direction of the low-pressure centre. If, for example, there is a small depression moving up the English Channel and the aircraft is above solid cloud somewhere over NW France, the chance of clearance will be increased to the east or SE. (Remember Ballot's Law on page 91 – when facing the wind, low pressure is away to the right, in the Northern Hemisphere.) It could be necessary to

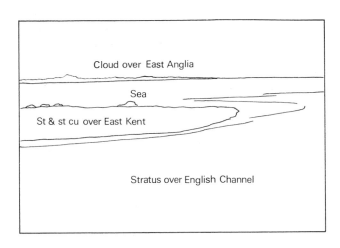

Cloud over East Anglia

Sea

St & st cu over East Kent

Stratus over English Channel

13.3 Although there is strato-cu over both east Kent and the Channel, the coastline itself is delineated by a marked slot. July 1971. Midday.

fly a couple of hundred miles, by which time frontier and other problems will be intruding, but when a depression is small, even though unpleasantly active, 40–50 miles in the right direction can produce a marked improvement. Even if no breaks in the sheet occur, cloud base is likely to be higher.

It is difficult, if lost above cloud, to find out which way the wind is blowing. If there are thunderheads sticking up through the sheet with the anvils leaned over by the wind, this will indicate that the wind is increasing with height and show the flow at that level. But it is still hard to decide the exact direction of the upper wind since there is nothing stationary by which to make reference. If the upper wind direction *is* known the chances are that the wind will be backed from this direction near the ground. Although guesses may seem a bit like clutching at straws, in the absence of any real information we have to turn detective and make the best use of what there is.

In little island countries the big risk when lost above cloud is of flying inadvertently out to sea, so a special watch should be kept on the cloud surface for any indication that this might be happening. As was seen earlier the coastline may be reproduced in the appearance of the cloud surface. If the wind is from the sea there will probably be some thickening of the cloud along, or a few miles inland of the coast, and it may have a more lumpy appearance over the land due to convection.

In clear air, if far from land, this may often be located initially by cumulus or other cloud over it. Even a tiny island will produce its private thermal topped by a white hat, or a whole coastline may be faithfully delineated by clouds. If the air is too dry for cumulus to form there will still probably be haze visible over the land.

It's better from below on tow . . .

Three weeks of bad weather is nothing unusual in December in Poland, but to have to cancel an urgent ferry flight flight twenty times and return by rail could be just too much. Finally it appeared to be OK and we took off on tow under a clear blue sky; that the small electric turn-and-slip did not work affected my happy mood very little. Suddenly, after half an hour, fog began to form underneath. As a glider pilot I am used to only seeing clouds from below, and I was not amused by the possibility of being separated from the ground by complete cover. I wanted to get down as quickly as possible, but my attempts to inform Stanislas, the

—continued

---

tug pilot, of my decision by manoeuvring the glider were unsuccessful. Steadily the high wings of the Gawron 4-seater towed me towards Warsaw 180 miles away, above complete overcast.

Alone in the tug Stan thought the situation would provide an interesting navigational exercise, and that the cloud below was local to the hilly country that we were crossing. He was still out of radio range of Warsaw, but talked to an airliner recently out of there which reported good visibility and 1/8 cloud. OK, in half an hour the ground would reappear.

The waviness of the cloud surface made it look as if a hole would show up just after the next cloud top; but there were no holes, and after an hour even Stan lost patience. Suddenly the Gawron lowered its nose, and with an altimeter reading of 1800 ft we dived into cloud. Slowly, at the rate of 6 ft per second we sank towards unknown ground. The spooky silhouette of the tug only 75 ft ahead became less visible, and then only barely discernible. Canopy icing! Open the vent. The outside world shrank to a narrow slit in the canopy, and maximum effort was needed to keep station. At 300 ft on the altimeter – it had been set at our airfield departure height – there was still no ground in sight, and after a short while I noticed that we were climbing again. I breathed a bit easier when we reappeared on top.

Stan finally got radio contact and requested weather and position. The chatter went something like this.

Air Traffic Control. Vis 1 mile, cloud 8/8, ceiling 700 ft, deteriorating. You are on course, proceed WA-NDB and report over Beacon.

Stan. Cannot proceed, have no radio compass.

ATC. What DO you have?

Stan (modestly). Altimeter, ASI, compass, turn-and-slip.

ATC. Stay on course, report in 20 minutes.

[Meanwhile visibility below had deteriorated to ½ mile and the ceiling to 400 ft.]

ATC. Gawron, course 15° right, descend to 2100 ft. You will be guided into the approach sector of the precision radar which will give you heading to steer on 121·5. Have you this frequency?

Stan. No.

ATC. (breathing deeply). Remain on 119·7. Radar will talk to you over the phone on this channel.

Stan (meekly). Request clearance for two approaches.

ATC. Say again! Why two?

Stan (resignedly). I am towing a glider.

ATC (very calmly). Stand-by, remain Visual.

[On the ground ATC tried and failed to find a clear alternate airfield, and realised that it would not only have to improvise, but change from international English to Polish to deal with the situation.]

—*continued*

ATC Radar. Gawron, I will guide you down. Steer 10° right. You are 5 miles from touchdown and 50 ft above glide path, descend faster. 100 ft above glide path, descend FASTER, 150 ft ab——
Stan (excited). Cannot descend faster, glider cannot follow.
Radar (unimpressed). 4 miles from touch down, descend faster or you will leave the glide path.

At about 900 ft it became dark, and despite my body contortions to see through the slit in the glider canopy, the tug disappeared from sight. This was lunacy, but to release would be even worse. Suddenly a large shadow loomed up – the Gawron was right in front. Airbrakes open! No sooner had it disappeared again than there was the first tearing jerk on the rope; but the second was more bearable. In the meantime my altimeter showed zero, then minus 60 ft. With the difference from departure height we should now be 500 ft up – I hoped not anywhere near the television masts.

Minus 150 ft. Still darkness
Minus 240 ft. Nothing to see
Minus 300 ft.
Minus 360 ft.

Suddenly the world appeared with industrial lights and a railway; still descending we shot past but I didn't read the station name. In a few seconds I landed. Nice to be back on Mother Earth, although with no airport building visible in the fog; only a small hut with two vibrating aerials – the precision radar.

—From an account by Adam Zientik

The concern in this chapter has been to help the pilot with a problem. But it must be emphasised that we have a responsibility not to lose control of the situation, since we are not the only users of the air. If, unsure of our position above cloud, we see a winged Polaris break the surface ahead, we must assume that we have strayed into controlled airspace. We should leave it quickly but sensibly, by flying away at right angles to the flight path of any airliner seen. If it is an airway 10–15 minutes flying, even in a slow aeroplane, should take us clear, but if it is a terminal area it will be more difficult to know which way to go, unless we can work out which one it is. If big jets are circling, we must think which stack it could be, and then leave in the direction which takes the aircraft out of the controlled airspace *quickest*, even if it is not our most convenient route. If the flight direction of big jets is noted, this may provide a clue to which airway has been entered, and help to sort out the navigation problem, *but under no circumstances should an uncontrolled let-down be made through cloud if there is the least suspicion that any part of this could be in controlled airspace.*

After any descent through cloud there is a chance that we will not immediately know where we are. Locating ourselves may be simply a matter of organising a study of the ground to orient the features which we see into the pattern in which they should be, in order to reach our destination; or it may be that we are quite lost. If the country seems entirely strange, the temptation to go blundering about in the vague hope of seeing some unmistakable feature such as the Forth Bridge or the Eiffel Tower should be resisted, particularly if visibility is poor, or cloudbase low. Firstly, we should lock on to any prominent ground feature and circle it, and go on doing so until we have made a plan. If the feature is a good sized town, patience, and the avoidance of self-deception, will in due course enable it to be identified on the map – provided that adequate map coverage has been brought along. If, however, the feature being orbited is of little navigational significance, the surrounding countryside should be studied until some other feature, which is obviously big enough to be represented on the map, is seen. The aircraft should be transferred to this feature and the map search begun again.

If lost in poor weather the temptation to continue on the original compass heading, or some insufficiently thought-out variation of it, is considerable. If the weather is deteriorating all that happens is that the chances of finding ourselves steadily diminish as the weather worsens – until finally the only possibility left is to land in some boggy or inconvenient field.

Study of the effect of ground features on the clouds may be made from the comfortable seat of a passenger jet. For some reason pilots often affect an air of boredom when travelling by airline and bury their heads in a book. This is crazy, when the acquisition of knowledge that may one day save our pride, if not something more valuable, is provided as a free supplement to the ticket.

Check list if lost in bad weather

1. If over the sea make for the *nearest* estimated point of land, taking into consideration the effect the wind will have on ground speed.

2. If over land stay with a prominent ground feature until you have calculated why you got lost, located your ground feature on the map, and made a plan.
3. Stick to the plan, and do not change it until a later properly thought out plan is made.
4. Check safety heights; those that will clear obstacles ahead.
5. In poor visibility reduce power and speed, go into fine pitch, and use partial flap – which usually gives a better view over the nose.
6. Keep checking fuel; if necessary and if possible weaken mixture.
7. Check that the compass is not affected by extraneous objects, such as camera light meters, etc.
8. Note times and bearings for all changes of course.
9. If irretrievably lost choose a really good large field and land in it while there is still at least 15–20 minutes fuel in the tank, and 20 minutes genuine daylight.

Weather accidents; take your pick

Starting off, or continuing to fly in weather which is more than we can cope with.

Failure to obtain a met briefing, either at all or one which properly covers the needs of the flight.

Misinterpretation of the forecast or briefing, either through ignorance or disbelief.

Failure to appreciate changes to the weather en route.

Observation of, but failure to do anything sensible about, deterioration in the weather.

Overconfidence in our ability to fly on instruments.

Overconfidence in our aircraft or its equipment, such as de-icers.

Personal reasons for pressing on with the flight regardless of the hazards.

Visibility may deteriorate rapidly and temporarily, due to a storm or a haze stream, or it may gradually worsen until it is no longer possible to use a map. This may happen under frontal cloud, and also in anticyclonic weather where the air under the inversion contains quantities of dust particles. If the inversion is only at about 1800–2000 ft, visibility will be extremely poor with all the suspended dirt compressed into this shallow layer. It may be impossible to see anything at all flying into the low sun during the early morning or late afternoon.

Anticyclonic haze is worst where there are big towns or industrial complexes, and when the calm subsiding centre of a High coincides with a smoke effluent area; visibility may drop to less than a mile even in summer. So in hazy weather our route should, if possible, be planned to pass upwind of industrial towns; even a 50-mile detour being worthwhile if it makes for easier map reading and a much reduced collision risk.

In poor visibility, particularly in summer, the aeroplane or helicopter pilot will be better off flying in the calm clear air just above the inversion, and out of reach of gliders milling round in the rough and murky thermals below. Map reading is often easier from above since visibility will be better when looking down vertically through the haze; also discoloration of the dust top by smoke from large chimneys can often be easily seen. From lower down, as we grope along in the murk, it can only be smelt.

A similar situation occurs with radiation fog (page 68). This develops as a result of the ground being cooled rapidly during the evening and at night. It is often confined to the lowest-lying areas such as river valleys, so that navigation is not usually a problem – it may even be made easier. Visibility is good above the fog with clear skies, and even if it is widespread there are usually enough features protruding through to provide frequent position checks. Streamers of smoke lying close to the ground or sea indicate that radiation fog may develop during the evening or night.

At night, or when the sun is below the level of radiation fog, it is possible to see, often clearly, vertically downwards through it, and the fog appears insignificant. Even with a vertical depth of 50 ft or so of fog it may seem as though there is nothing there at all, but on the approach we find ourselves suddenly blind. This is because we are now flying, and looking ahead, through perhaps 1500 ft of fog, instead of 50 ft, and we cannot see anything at all; landing lights will only worsen our predicament. If fog is suspected a low, but safe, pass should be made over the field to discover the height of the top of the fog. If there is no alternative to making a landing at this place, the aircraft should be climbed and flown around over the airfield until the layout and the position of any obstructions are thoroughly familiar. We should then come in on a steep, slow, power-on approach. With radiation fog there will be almost no wind or turbulence so that whichever run offers the safest line could be used. It will probably be only the last 50 ft or so above the ground that will be thick, so if the aircraft is properly lined up it should not be too difficult. If snow is lying it should be expected that runway landing lights may be partly submerged and therefore not be so easy to see from a flat approach angle. They may even disappear altogether during the round-out.

If the sun is above the level of the radiation fog, as is likely in the morning before it has dispersed, it is less easy to see downwards, because of the reflections from the top of the layer, and it will be just as difficult to approach safely through it. There will also be an eyeball problem in the sudden change from bright sunshine to gloom. After sunrise the new warmth of the day starts to stir up the air, so the radiation fog will often thicken up for a time before dispersing; it may then just burn off by evaporation, or lift and change into strato-cumulus. The aircraft fuel supply will probably determine whether we will be prepared to wait in the vicinity for the fog to clear, or divert to another airfield; information from the airfield locals on the expected time of clearance is likely to be reliable.

Radiation fog occurs earliest and most extensively, and lasts longest, in the lowest-lying areas, especially saucer-shaped valleys with no air-draining outlet at the end of a main valley. Diversion airfields should be chosen from those which are at a greater height, or on a slope, or possibly on the coast where there is air flowing from the land towards the sea. Other valley airfields on the same day should also be suspect; since, for obvious reasons, airfields are built in the centres of valleys they may be blotted out by a thin layer of fog even when the slopes are clear.

Occasionally a low-lying cold air pool in a deep valley is at such

A Christmas Story

... 34 minutes to go before sunset. We would not make it by nightfall, twilight is brief in the mountains. Over Donner Summit the snow sparkled in the dusk. 'No moon this flight,' commented my wife. I called the airfield and they came back immediately, loud and clear: wind calm, Runway 28 in use, some fog, aircraft taxiing out for take off. 'Roger,' I replied, 'have him in sight.' What I did not have in sight was any hint of fog. The valley and 6000 ft of lighted runway lay unfolded before us, mantled in the silent snow, crisp and cool at 5900 ft. We completed our checklist, then gazed down at the unblinking lights as we circled the field. Everything was OK, runway lights clearly visible, two notches of flap. As the approach progressed, I became suspicious that the landing light was inoperative, but we continued the approach. I chopped the throttle just before we came over the threshold. Normal flare. Then the beautiful trap sprung hard. Everything outside the cockpit went grey. We continued to settle a few sickening seconds; there was a hard landing, a brief ground roll, and then a terrible lurch with the loud hollow report of structural destruction. 'Daddy, why did you do *that*?' my five year old shouted.

Presently I sized up the situation, decided that no gasoline was spilled, observed one wing partly overhanging the runway, and turned the master switch on again. There followed a threeway radio conversation which included an Aztec taking off. He couldn't see us, nor we him. Presently it roared past at take off power, trusting my estimate of clearance from my overhanging wing.

Before we could get a warning flasher out on the aircraft an Aero Commander called in. He could see the airport clearly from above; then 'Lost runway on final.' A little later he made a second pass and taxied in. By noon the next day the fog was still there. Nonetheless, several aircraft announced their intention to land, and the trap continued to sing its siren song. The pilots protested that they saw the runway well; they could be heard over the field, but they appeared in view for only a moment, directly overhead. Except for the noise of their engines, shrouds of mist maintained silence over the ground.

—From an account by Albert I. Harbury, *AOPA Pilot*

a frigid temperature that air which could otherwise flow down the hillsides and into the area and 'flush it out' simply moves across the top of it. The cold air, fog or frost, is trapped and may remain so for days.

Sea fog is caused by condensation in moist air flowing over cold water, or sometimes when cold air flows over warmer water or coast. Usually fog disperses as it moves in over broken or warmer land, but

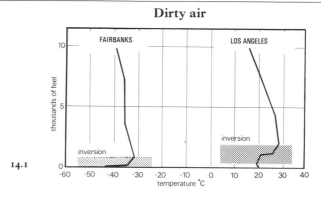

Dirty air

The main types of air pollution are:

(a) coal smoke and gases, still the major offender;
(b) various toxins, usually related to industrial waste, and localised;
(c) smog, the result of photo-chemical reactions involving unburnt hydrocarbons, mainly in urban areas (e.g., Los Angeles);
(d) ice fog, the result of the inability of really cold air to dissolve the water vapour effluent of an urban environment.

An example of (d) occurs at Fairbanks, Alaska. When the temperature goes below $-35°C$, as it frequently does, the water vapour pouring out of cooling towers, car exhausts, people and heating plants, is not absorbed, but crystallised directly into tiny ice crystals, spherical or columnar in shape and about 5 micrometres in diameter. Some 30,000 kg of this moisture comes from the breathing alone of the 30,000 people and 2000 dogs living there.

In winter, the Alaskan air is cold, stable, and calm for long periods, with a strong inversion of up to $3°C$ per 100 metres starting at ground level. Fairbanks lies in a stagnant air basin, with the midwinter sun achieving just $2°$ above the horizon; so there is practically nothing to disturb the increasing quantity of dirty ice dust in the surface layers. The crystals have a large surface area and falling at less than 1 cm/sec easily absorb any pollutants present. The depth of the ice fog may not exceed 60 ft, but it is enough to reduce street visibility to less than 15 ft, and these conditions may persist for days on end.

Most cities produce enough heat to weaken, or break in due course, inversions above them, but although Fairbanks increases the local air temperature by some $6°C$, this is not enough to destroy the inversion.

Although the smog of Los Angeles is bad enough, its dispersal is aided by the air layer below the inversion in which there will be some turbulent mixing and wind; also the much higher temperatures enable more moisture effluent to be absorbed.

if this is flat and cold it may penetrate some distance. Often the mixing of the air produced by the rough surface of the land lifts the fog into low stratus, or ragged strato-cumulus.

In summer sea fog tends to be very local, although sometimes it persists until well into afternoon, but in winter it may last all day, particularly if there is also high cloud to reduce such warmth as the winter sun provides. Sea fog gives us a serious problem only when there is insufficient fuel to go elsewhere, and this might only have to be a few miles up the coast.

14.2 Sea fog drifting in. Falmouth, August.

Visibility

Visibility descriptions have the following meanings for (a) sea, (b) land, and (c) coastal stations respectively.

GOOD	(a, b, c) more than 5 nautical miles
MODERATE	(a, b, c) 2–5 nautical miles
POOR	(a, b, c) 1100 yd to 2 nautical miles
FOG	(a, c) less than 1100 yd; (b) less than 200 yd
MIST or HAZE	(a, b) 200–1100 yd; (c) 1100–2200 yd
DENSE FOG	(a, b, c) less than 50 yd.

The term 'fair' is given in forecasts when there is nothing significant about the weather. It may or may not be cloudy.

15 Winds

Some countries, like Britain, Iceland or New Zealand, have a good share of the world's winds, often more than seems necessary. Others like Switzerland, in the middle of mountains and within a large land mass, give the impression of being quite frequently windless. There is obviously some substance in this since the Swiss will pin up, on outside walls, thin paper notices just with drawing pins, whereas in Britain it is necessary to paste them firmly if they are to survive. On weather coasts, such as the Atlantic face of the British Isles, the wind blows so continuously that a calm day comes almost as a surprise. When landing at a coastal airfield the wind may be stronger than it was inland, with a local direction 'bent' by the line of the coast.

Since wind can substantially alter our fuel range we should question the met man closely on this matter if the flight is to be a long one. In addition frequent checks during the flight should be made by radio, or by eye, to discover whether the wind is remaining as forecast.

From the air there are several ways of checking the wind. At cumulus level the direction of the wind can be discovered by watching the passage of their shadows over the ground. Some prominent line, such as a road or railway the direction of which is known, should be selected, approximately the same as the wind is thought to be; any cloud shadow near this line should be observed for convergence or divergence, and its direction determined. The speed of the shadows over the ground can be guessed if compared with that of trains or cars – if the shadow starts overtaking cars on a motorway, the pilot's alarm bells should trigger off.

If there are no clouds a few continuous air circles, starting from a known point such as a large house or lake, will indicate the drift direction, unless the wind is very light. The drift of the aircraft's shadow when flown low over ground or water will, of course, give an idea of the wind near the surface, but it is not always practicable or lawful to fly low enough to do this effectively.

When landing it is only the surface wind which is important, so indicators on the ground, such as smoke, ripples of water, or tall crops in open areas should be used. Flags, and washing hanging

A

15.1 Cirrus

These photographs show how cirrus looks in a strong wind.

The hooked shape of the cirrus in (A) is caused by a rapid change of wind speed with height.

In (B and C) there is strong wind cirrus blowing from different directions at adajcent levels. Obviously there is a lot of action taking place at these levels, and the forthcoming weather is very likely to be equally disturbed.

B

C

out, are usually situated in the turbulence of buildings and are less reliable. Wind lines on the sea can usually be easily seen, but should not be confused with wave swell, which may be running in a different direction (page 240). The wind direction may, for various reasons, change during the course of a flight. In big air mass systems – depressions and anticyclones – it will change progressively at a given spot as the weather system moves across it. The wind will also change as far as an aircraft is concerned, as the aircraft moves

15.2

across the system. If this is small there may be considerable change in wind direction on a flight of only 100 miles or so (Fig. 15.2).

Sometimes, for no obvious reason, the wind becomes stronger and stronger. Following the passage of an active trough, cloud clears, but the wind, instead of continuing to blow at, say, 15–20 knots, goes on increasing. This may occur because the trough is still intensifying and accelerating, or because there is a squeeze effect between the Low and an adjacent anticyclone. If the sky stays clear and the flight is being made at 5000 ft or so, above the turbulence, and without regular position checks being made, there may be no warning signs of the wind increase. The height at which the aircraft is flying will be too great for ripples in crops to be noticeable, although over open water white horses and extensively blown spume from waves can be clearly seen from as high as 30,000 ft. On such occasions the wind may increase to 40 knots or more in the lower levels, particularly along the coast. Flying against winds of this strength may bring fuel problems but even if they do not we could have a landing one, with powerful wind gradients and really rough

---

The pilot said . . .

. . . he was unable to locate the windsock so followed another aeroplane in. As he touched down he bounced, veered off the runway and went on his nose. There was a 90° crosswind of 15–20 knots.

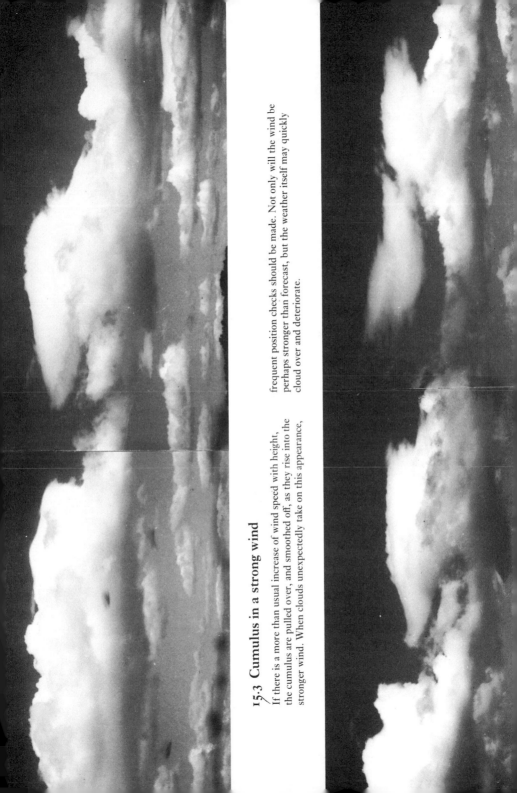

15·3 Cumulus in a strong wind

If there is a more than usual increase of wind speed with height, the cumulus are pulled over, and smoothed off, as they rise into the stronger wind. When clouds unexpectedly take on this appearance, frequent position checks should be made. Not only will the wind be perhaps stronger than forecast, but the weather itself may quickly cloud over and deteriorate.

air. If it is realised early enough that the wind has become too strong, the best course is to divert, preferably to an inland airfield free of surrounding hills and other obstructions.

Wind gradient

This is the slackening, often abruptly, of the wind near the surface due to the drag caused by the ground and its obstructions. The gradient will be least over a smooth open airfield or water, and sharpest where the landing area is surrounded by broken ground or buildings. The effect of trees or hangars upwind of the touch-down point may be considerable. If there are hills nearby there may be areas of dead air close to the lower slopes, or even a reversal of direction due to eddies or curlover. Flying slowly through such ground turbulence or powerful wind gradient is like asking to be dumped on the doorstep.

During an approach the wind strength can be checked from the sock, and also from the amount of drift found on the base leg. If it is considerable an extra margin of speed should be used on finals. This is particularly necessary for slow, lightly loaded, or long span, aircraft. If the air seems very rough steep turns, which will put the two wings into winds of appreciably different speeds, should be avoided. If, after landing, there is difficulty in taxiing it is better to continue slowly directly into wind to clear the landing area, and then wait with the engine running until help arrives. If none is forthcoming, the slow into wind taxiing should be continued until some windbreak, even an openwork airfield fence, is reached. The aircraft should then be tied down as quickly as possible. If solo the stick should be jammed forward before getting out of a glider or some lightly loaded aeroplanes. In addition to the usual picketing points the aircraft should be additionally secured by the wheels and any other suitable attachment points if the wind is near gale force; one owner found only the ropes and the picketing rings left the following morning. A blow-over may often be avoided by parking a car just upwind of the aircraft. Often a small windbreak is more effective than parking to the lee of a large building round which severely turbulent air is sweeping.

Whenever tying down an aircraft in strong winds, the direction in which the wind is likely to change should be considered; if a new depression is whistling in, the wind is likely to back; if a cold front is going through, it will veer. In mountainous country which

A

15.4 Cumulo-nimbus in a strong wind

A The ice-crystal 'anvil' of this cu-nb grew at great speed into the very strong winds that were blowing at greater heights. It appears to have passed through a slightly warmer and perhaps slower moving layer on the way.

B This cu-nb anvil grew up into the stronger winds. The air at these heights was also more stable causing the head to flatten out and collapse quite quickly.
25 July. 1255 and 1310 hrs. Surface wind W (L to R), 18 knots.

B

> ### 'My instructor said, if in doubt, don't. But I did ...'
>
> For two weeks I had been looking forward to my first long solo cross country. Off I sailed, dumb and happy, towards Piqua, 118 miles away. On entering the pattern to land, I noticed the following breeze was no longer a breeze, but a wind. I landed OK, but when I turned on to the taxiway I discovered that the engine was no longer needed to taxi. The wind pushed the little Champ along with no trouble at all. In fact, when I shut down the engine and opened the door to get out, the Champ started moving out over the boondocks all by itself. I hit the brakes, but could not even get out until the line boy had chocked the roving Champ. I gassed up, checked the weather because there was now a good overcast, and inched my way homewards against the wind. As I pushed west, the overcast became thicker and darker, but because of my nonexistent experience with weather I pushed on, barely maintaining VFR. About half way I spotted Randolph County Airport and almost decided to land; but didn't. This was a big mistake. Now the wind was really strong. My ground speed was down to 30 mph, and I was crabbing along in slight rain. But I was still VFR and the visibility was decent. But the Champ had turned into a wild animal bent on bouncing me all over the sky. It was taking nearly everything I had to keep on course, and right side up. I was only a few minutes out of Reese, where I had intended making a landing, when I saw one of the most unnerving sights of my life. Stretching from the SW–NW was what appeared to be a black curtain hanging from just above my altitude all the way to the deck. I looked behind me and found that the weather had closed in, and I was trapped. I couldn't turn around and go back. I had to land at Reese, and I had to do it quickly because the wall of cloud appeared to be closing fast.
>
> The wind was now between 30 and 35 knots, with gusts up to 50, and shifting madly. It wasn't down any runway. I turned final, and it took a considerable amount of power to keep moving when the wind was against me. It shifted crosswind and blew me away from the runway. I went around and tried again. It took plenty of power to hold the plane over the runway, but somehow I got it to within a few feet of the ground. I hit very hard, bounced, and fought my way back to the ground again. I rolled a few feet, skidding sideways all the way, and as soon as I stopped men were on the wing tips holding on for dear life.
>
> —From an account by David Quick when a student pilot, *AOPA Pilot*.

is producing waves, the wind at surface will be augmented where the downgoing part of the wave reaches the ground. In a gale force wind of 50 knots, the addition of the wave may locally push the speed up to 70–90 knots. This is almost certainly what happened in recent years when severe damage was caused in a part of Sheffield, and when many acres of a Scottish forest were destroyed. Both disasters happened at night. In such winds no aircraft, however well parked or even hangared, is absolutely safe.

We know that hills and mountains will always upset the air, but flying over hilly country even in quite strong winds rarely produces a problem provided that we know the direction of flow and can keep the aircraft away from downcurrents and the severe turbulence that often accompanies them. Over a simple ridge the wind flow is usually obvious, but in broken mountainous country it may be impossible to find out in which direction the wind is blowing locally at low levels. Even experienced glider pilots, studying the situation intimately in order to stay airborne, may sometimes get it wrong; there may be just too little visual evidence available to decipher what is a complex aerodynamic problem. If intending to fly through a mountainous region at no very great altitude in strong winds, the answer is don't; but if there is no alternative the aircraft should be kept near smoother wind-facing slopes. When trying to escape from savage turbulence it is best to leave the area along any valley running up and down wind, and not by those running across wind.

Some countries have to perch their airfields and strips on the tops of hills, because that is the shape of their terrain, so if the into-wind approach is over surrounding low ground towards the cliff or hill, downcurrents must be expected on the way in. The locals will have discovered that there is a certain approach angle above which conditions will be entirely normal, but below which there will be both downwash and turbulence. If we fortuitously come in above the critical angle, we may disbelieve the rumours of problems that we have heard about. If, however, on our next approach, we come in low, and perhaps also in windier weather, we will find ourselves sinking rapidly in exceptionally rough air, with the airfield becoming increasingly out of reach above us. Nearer the hill face there may be dead air or a reversal of the flow, so that the headwind suddenly changes to a blast on the tail, accelerating the aircraft towards the hill with declining airspeed. The dilemma becomes considerable if the power of the motor is inadequate to gain, or even maintain, height. So if on an approach to a clifftop airfield there is any doubt that the desired approach path can be maintained, the aircraft should be turned away, with the nose really well down, power on, *before it is too late*. As a general

		Wind	
The Beaufort Scale			
Beaufort Force	*Description*	*Effects on land*	*Wind speed/knots*
0	Calm	Smoke rises vertically.	Less than 1
1	Light air	Wind direction shown by smoke drift, but not strong enough to turn wind vanes.	1–3
2	Light breeze	Wind felt on face. Leaves rustle. Wind vanes respond to wind.	4–6
3	Gentle breeze	Leaves and small twigs in constant motion. Wind extends light flags.	7–10
4	Moderate breeze	Raises dust and paper. Moves small branches.	11–15
5	Fresh breeze	Small trees begin to sway.	16–21
6	Strong breeze	Large branches in motion. Telegraph wires hum. Umbrellas difficult.	22–27
7	Near gale	Larger trees sway. Inconvenience felt when walking against the wind.	28–33
8	Gale	Breaks twigs off trees. Generally impedes progress.	34–40
9	Strong gale	Slight structural damage to chimney pots, slates, TV aerials, fences, etc.	41–47
10	Storm	Trees uprooted.	48–55

rule approaches into hilltop fields from over the valley should be made from a steep glide path, with the landing made as far up the field as is safe.

If the take-off is towards and over the edge of the hill, a high climb-out should be made if this edge is abrupt like a cliff, since there will be a dead area just over the lip; but if the fall away is from a smooth rounded hill, the aircraft can be taken out into wind safely at any height. If the airfield is a few thousand feet up the take off run will be longer than at sea level. If the day is also hot and/or there is an upslope, long grass, any tail wind, or a full load the airfield may not be long enough. (See Appendix 3, Density Altitude.)

The Föhn

The Föhn is a well-known wind created by mountains whose effect on the local weather is considerable. We know that air flowing towards mountains is pushed up, and if it is moist the processes of condensation will soon produce cloud. If there is further upward movement this will result in precipitation either of rain or snow,

15.5 The Föhn wind

The Föhn is a warm dry wind which blows to the lee of mountains. The air rises on the windward side and cools at the dry adiabatic lapse rate of 3°C per 1000 ft up to cloud base, and at the saturated rate of 1·5°C in cloud. It warms at the same rates on its way down the lee slopes, but is warming at 3°C over a greater height range.

which falls on the mountains. This precipitation is moisture lost from the air, so that which flows on over the lee slopes holds only such moisture as remains suspended as cloud or as water vapour. The air is now drier than when it rose on the windward side of the mountains, and so will return to an unsaturated condition at a greater height on the lee side, and will warm adiabatically during the descent (Fig. 15.5) As well as dry and warm, the Föhn wind may be strong, so it will have a dehydrating effect on the surface. It can, for example, evaporate as much snow in a day as the sun alone is able to evaporate in two weeks, so it is regarded with dislike in ski resorts. I have seen ski slopes turn to porridge within a few hours in an astonishing mid-winter temperature of 10°C at 7000 ft in the Alps. The Föhn may also create a considerable fire risk as it dries out grass and pine forests. Before Swiss houses were made of other than wood whole villages were burnt up as the result of the Föhn. Glarus was destroyed in 1299, 1337, 1477, and again in 1861, the strong wind blowing debris over a wide area and setting fire to anything combustible in its path. In Switzerland alone the Föhn blows on about 40 days a year mainly over the north face of the Alps, but it is also common in such places as Wyoming and Montana, USA, especially in winter. It is most likely to occur ahead of an approaching depression. During its crossing of the mountains air which becomes the Föhn wind tends to change direction, blowing more across the isobars at lower levels and towards the main low-pressure centre. If there is extensive cloud over the mountains and in the air which is cascading down the slopes, this will evaporate on reaching the level at which the increasing temperature allows the air to become unsaturated. Looking up at the cloud from the valley it appears as a stationary wall covering the mountain top, and is known, appropriately, as the Föhn wall.

Flying in Föhn conditions will be better in the warm clearer air to the lee of the high ground, than in the moist and cloud-covered windward side, although turbulence and downcurrents should be expected in the strongly descending air close to the mountain itself. The Föhn situation is conducive to the formation of waves, so may give large surges of rising and sinking air for many miles downwind.

If the warm Föhn wind encounters stagnant pools of cold air lying in sheltered valleys it may simply flow over this cooler air (page 228). Where this happens not only the temperature but the visibility and any wind will be different in the two lots of air, so

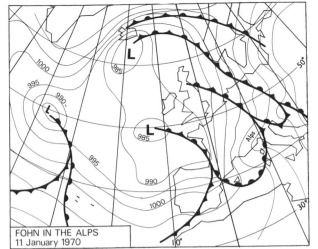

FOHN IN THE ALPS
11 January 1970

A

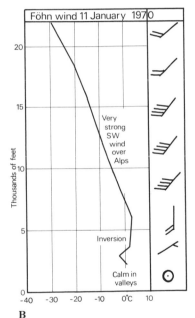

Föhn wind 11 January 1970

Very
strong
SW
wind
over
Alps

Inversion

Calm in
valleys

Thousands of feet

B

15.6 Waves in the Föhn

The Föhn may blow across the Alps some 40 days a year, as a soft and gentle breeze or as a powerful torrent of warm dry air.

The 'starke Föhnwind' provides the glider pilot with lift to more than 20,000 ft.

On 10/11 January 1970, while snow in the Swiss Engadine was turning to porridge, gliders were soaring to 31,000 ft in places as far apart as Arosa and the Grossglockner. While skiers were sweating it out in temperatures of 10°C in Davos, pilots were soaring in temperatures of − 30°C.

The synoptic situation which produces the Föhn over the Alps comes from a low pressure system with SW winds up to considerable heights. The chart (A) shows the surface and (C) 500 mb pressure distribution for 11 January 1970.

Before crossing the Alps when the Föhn is blowing hard, weather information should be obtained from *local* met men and airfield operators. They know which valleys will remain clear and for how long.

C

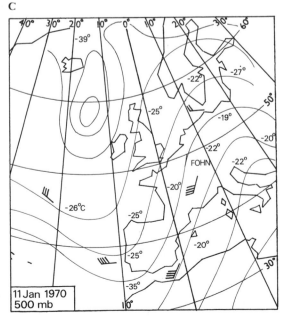

11 Jan 1970
500 mb

if our destination airfield is located in such a cold air pool such changes should be expected during the approach.

Although the full effect of the Föhn is only obvious when it blows over fairly large mountains, to a lesser extent there will be warmer and drier air to the lee of almost any high ground. The Pennines are an example; in a westerly wind a great deal of moisture is deposited on this range of hills so that the flat lands to the east, and the air over them, is drier.

Valley winds

Valleys, even large ones, invariably have the wind blowing in one of two directions – up the valleys, or down them – even when at higher levels the main wind is blowing across. Most valley winds are caused by convection. During the day the sunny sides of moun-

15.7A Anabatic wind
Air flows up the mountain and thermals rise from the top.

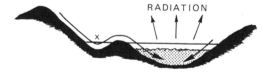

15.7B Katabatic wind
Cold air flows down the mountain. Where the air is cooled below the dew point fog forms. × shows a typical frost hollow, from which cold air cannot drain.

tains are warmed and the air rises up the slopes. This is called an *Anabatic* wind. Thermals continue to rise above the mountain tops and air is drawn UP the valley towards the lower pressure caused by the rising air. In the evening the mountains cool while air down in the valley remains warmer longer; air now rises from the middle of the valleys with a return flow down the mountain sides. This is a *Katabatic* wind. The general flow of air will now be DOWN the valley. At night and in winter unless there is a moderate to strong general wind, air in the valleys is often calm and cold.

Some valleys, due to a combination of meteorological and topographical features have local winds of quite remarkable strength, enough to cause the whole region of the valley to become inhospitable to the itinerant small aeroplane. Such a situation occurs in the main Rhone valley of France, home of the notorious Mistral.

The Southerly Buster

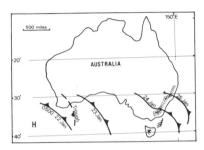

15.8

The southerly Buster is Australia's famous wind. It is really an extremely active squall line, and blasts up the New South Wales coast accompanied by fierce gusts, quantities of dust, and thunder, lightning and rain. Its appearance is dramatic; a dark roll of cloud up to 60 miles in length comes up over the southern horizon, with low scud forming underneath and being swept up into the rotating blackness. As the storm arrives the wind first drops and then swings violently to the south, blowing at gale force.

The Southerly Buster is born $1\frac{1}{2}$ thousand miles away on the west coast of Australia as a simple cold front, sometimes accompanied by an airmass trough, lying between two eastward-moving anticyclones. The front steadily accelerates as it moves across the southern coastline towards the well-heated land mass of Victoria and New South Wales. Here it meets warm air which is also moist. The cool frontal air thrusts under this warm air and powerful convection develops along the line. The front swings round and up the coast of New South Wales, due partly to the drag effect of the high ground, and partly to the Bass Straight, which offers no impediment to the wind. The speed of the front is now slowed down, but this has little effect on the strength and gustiness of the local wind.

After about half an hour of turmoil, the violence of the front passes, and a cool southerly flow becomes established.

This wind, which may last for several days, is a strong cold northerly blast which blows down the valley for 100 miles. Air is funnelled between the Alps and the Massif Central, blowing with exceptional strength down the relatively narrow valley that lies between them. This northerly flow of air also sets off waves over the adjacent mountains which may go higher than 20,000 ft. Unfortunately for the glider pilot who tries to use them and fails, there is a shortage of good landing places in parts of this region.

Similar winds to the Mistral include the Bora, on the Dalmatian coast, and the Buran, the cold north-easter which blasts across Siberia. Even if we are not forewarned about special winds when planning to fly across some dubious mountain area, we should be able to size up the situation for ourselves. We can study the shape of the mountains from the map, build in typical wind directions and strengths, and work out what the air is most likely to do. If the wind is pouring over the top of a mountain range it will channel itself into suitably shaped available valleys just as water does, and it will flow in a turbulent manner over foothills. The effect of centuries of wind can be easily seen on some mountain slopes which have a scoured, almost streamlined appearance. In particular, cold air downfalls may produce very strong winds because the air is dense and wants to go down anyway.

Notorious winds			
Name	Location	Season	Description
Belat	S. coast Arabia	Winter	N–NW. Strong land wind
Bize	France, Southern Mts	Winter	N. Cold, dry, cloudy
Bora	Dalmatian Coast	Winter	NE. Cold, gale
Brickfielder	Australia	Summer	N. Hot, dusty, from interior
Buran	Siberia	Winter	NE. Cold, with blizzards
Chinook	N. America	Anytime	Föhn wind
Elephanta	India, Malabar	End of Monsoon	S–SE. Gale
Etesian	Greece, Aegean	Summer	N. Fine, clear
Föhn	Alps	Anytime	Warm, dry, lee of mountains
Gregale	Malta area	Winter	NE. Gale, squally, persistent
Haboob	Egypt	Anytime	Sandstorm
Harmattan	Cape Verde, W. Africa	Nov.–March	ENE. Hot, dry, squally, from desert
Helm	Lake District	Anytime	E. Lee wave wind
Kachchan	Ceylon	Anytime	Föhn wind
Khamsin	Egypt, Aden	Spring, summer	From Sahara. Hot, dry, some sandstorms
Kaus	Persian Gulf	Winter	SE
Kharif	Berbera, Aden	SW Monsoon	SW. Strong, diurnal

Name	Location	Season	Description
Leste	Madeira	Anytime	S. Hot, dry, ahead of depression
Levantades	Spain, E. coast	Spring, autumn	N. Gale
Levanter	Gibraltar Straits	Summer, March	E. Hot, damp, not strong
Leveche	Spain, SE coast	Summer	SW–SE. Hot, local, precedes Low
Libeccio	N. Corsica	Anytime	W–SW. Squally
Maestro	Adriatic	Summer	NW. Fresh, fine, sunny
Marin	S. France	Anytime	SE. Warm, moist, cloudy
Meltemi	Aegean	Summer	NW–NE. Fine, clear
Mistral	Rhone, France	Anytime	N. Frequent, cold, gale, often clear
Norther	Panama, Mexico	Winter	N. Cold, dry, strong
Pampero	Argentina	Winter	SW. Gale with line squall
Papagayo	Costa Rica	Winter	N. Cold, dry, strong
Poorgas	Tundras	Winter	NE. Cold with blizzards
Santa Ana	S. California	Winter	E. Hot, dusty, some Föhn effect
Shamal	Persian Gulf	Summer	NW. Hot, dusty, cloudless, nights calm
Seistan	E. Persia	Summer	N. Strong, continues 4 months
Simoom	Palestine, Syria	Summer, autumn	SE–E. Hot, dry, dusty
Sirocco	Malta, Italy	Late summer	SE. Hot, damp, ahead of depression
Solano	Gibraltar, Spain	Anytime	SE. Wet
Suahili	Persian Gulf	Winter	SW. Very strong, follows Kaus
Sumatra	Malacca Straits	SW Monsoon	Squally, thundery, wet
Southerly Buster	Australia	Anytime	S. Line squall in trough, cool
Tehuantepecer	Gulf of Tehuantepec	Anytime	N. Cold, dry, strong
Tramontana	N. Corsica, Cyprus	Winter	N–NE. Fresh, fine weather
Williwaws	Magellan Straits	Anytime	Heavy squalls
Vendavales	Gibraltar, E. Spain	Spring, autumn	SW. Strong, squalls, thunder
Zonda	Argentina	Anytime	Föhn wind

16 Wild winds

When we think that wind is only air moving about to equalise pressure in a free atmosphere, it seems surprising that it can sometimes get so strong; a wind of 80 mph blowing over a distance of as little as 50 miles for perhaps only an hour is shifting an awful lot of air. But it does happen, and some of the more extreme manifestations of wind are, of course, the Hurricane, Typhoon, and Tornado – good names also for aeroplanes and cars. All these winds are caused by large and often local falls in pressure, so air starts moving in and circulating around, just as it does in an ordinary depression. The difference is that the pressure gradient – the fall of pressure towards the centre – becomes really steep; as the central pressure plummets, so the winds really whistle round the eye, as it is called.

The Hurricane, Typhoon, and Cyclone are the same, except that the hurricane comes from the Caribbean and the others from Asian seas. They cause severe damage, not only by their gale-strength winds and rapidly sinking pressure, but by associated confusions such as tidal bores – millions of tons of sea being piled up and swept over any land that happens to be in the way. The only reasonable thing about hurricanes and typhoons is that they usually have a short season, that of late summer, when the sea is at its warmest and the sun is still hot. They develop quickly, with rapidly dropping pressure and produce cloud up to great heights. The centre often remains clear and calm. They travel similarly in direction to the more mundane depression; a Caribbean-spawned

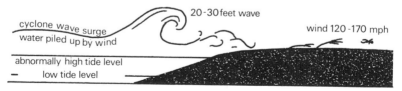

20-30 feet wave

cyclone wave surge
water piled up by wind

wind 120-170 mph

abnormally high tide level
— low tide level

A

16.1 The Bay of Bengal regularly has six cyclones each year all causing some loss of life. Once or twice every 100 years a super cyclone hits the coast with devastating results. In 1737 300,000 people died and in 1876 100,000. The most likely period is October–November.

B

16.1 The two satellite photographs (B and C), taken 25 hours apart,
show the progress of the cyclone of 12 November 1970 which
caused many thousand casualties.

C

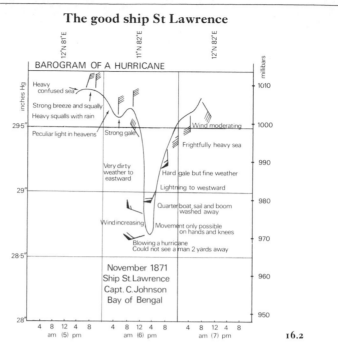

The good ship St Lawrence

12°N 81°E   11°N 82°E   12°N 82°E

BAROGRAM OF A HURRICANE

inches Hg

Heavy confused sea

Strong breeze and squally

Heavy squalls with rain

29·5″

Peculiar light in heavens   Strong gale

Wind moderating

Frightfully heavy sea

Very dirty weather to eastward   Hard gale but fine weather

29″

Lightning to westward

Quarter boat, sail and boom washed away

Wind increasing   Movement only possible on hands and knees

Blowing a hurricane
Could not see a man 2 yards away

28·5″

November 1871
Ship St.Lawrence
Capt. C.Johnson
Bay of Bengal

28″

4   8   12   4   8   4   8   12   4   8   4   8   12   4   8
am (5) pm   am (6) pm   am (7) pm

millibars

1010

1000

990

980

970

960

950

16.2

In November 1871 the sailing ship St Lawrence was bound for Madras from London, and extracts from the log of the Captain give a good idea of the weather in a cyclone in the Bay of Bengal.

'Sunday, 5th, 6 pm, beginning to look dirty with a heavy confused sea. Monday 6th, 1 am, sea heavy. Very heavy squalls with heavy rain. A peculiar light in the heavens like light brick dust. 3 am, stowed upper topsails. Strong gale NNE. 6 am, wind unsteady from NW to NNE, heavy squalls, glass rising a little. 10 am, down fore and mizzen topgallant sails. Fearfully heavy sea. A paddy bird and flight of curlews tried to settle, but could not. Noon, blowing a perfect hurricane and such a tremendous sea. I was afraid to run any longer having had the cuddy, maindeck, and forecastle filled. Hove to under bare poles with a small boat sail in mizzen rigging. Wind NW. 1.30 pm, mizzen stay parted. Increasing every minute; could not get along deck except on hands and knees. Ship lay-to beautifully with water just washing over the lee poop. 2 pm, blowing a furious hurricane. Wind WNW. Starboard quarter boat washed away, also starboard mainrail and hammock nettings. Jib-boom and fore-topgallant mast gone. 3 pm, still blowing a romping, roaring vicious hurricane. How the masts stand I can't think. 3.30 pm, glass commenced to rise. Wind W. 5 pm, wind SW but still blowing a thunder-

—continued

ing hard gale; obliged to cut away the wreck of jib-boom to save the ship's bows. A great deal of lightning to the westward, but getting fainter. 'Tuesday, 7th, 4 am. Frightfully heavy sea. During the hurricane observed a peculiar glare in the heavens when there was a lull, but for nearly three hours nothing could be seen on board, and our eyes and ears suffered for it afterwards. Could not see a man two yards from us.'

The pressure drop met with by Captain Johnson is equivalent to an altimeter error – the altimeter showing height that does not exist – of 1000 ft.

hurricane, for example, moving generally N and E. Their track may sometimes take them over land, such as Florida. If it does, damage will be worst near the coast because over the drier and more broken surface further inland the hurricane will soon lose its power. In their developing phase, however, young hurricanes may travel in a westward direction before turning N and E, and in this respect are a give-away to the vigilant forecaster.

Hurricanes, which are now identified by assorted female names, are not often more than 100 miles across, although their turbulence will affect a much greater distance. Before the structure of such storms had been scientifically explored, sailing ships were frequently sunk by being misled by the calm eye. Having survived the storm, or so they thought, the ship would suddenly be hit by violent winds from the opposite direction. With sails aback they would be swamped and sink by the stern. The greatest aid now to warning systems is the satellite watch which enables a continuous check to be made on the birth and angry life of every hurricane and typhoon, so no longer is it necessary to get caught unawares. This is not the case, however, with the smaller and more local tornado.

Some people call this vicious whirlwind a twister. It also requires a plentiful supply of moist air to grow properly, but it is usually associated with a bigger storm or squall line. A particularly powerful updraught in the squall sets up a small spiralling depression which lowers the pressure locally. The circulation around the centre is very fast and the ascending speed high; the high rate of spin being due to the increase of vorticity occurring when air converges towards the updraught. Condensation takes place in this updraught, in which air is cooling so rapidly by expansion that the cloud appears to grow downwards. Such swirling tails of cloud can often be seen below the dark base of a storm, and they are accompanied by really fierce winds. The damage caused along the narrow tornado track is achieved by the wind and, even more, by the 'suck' of the sharp pressure drop. This causes buildings to explode unless doors and windows have been left open. At sea the downward tongue of spiralling cloud may reach the surface as a water spout. As condensation takes place rapidly down the low pressure 'tube', it is often met by a sort of stalagmite of sea drawn up to join it.

Tornadoes develop rapidly, but because they are associated with a bigger cloud mass they may not be noticed in time to get out of the way. A small alteration of course to go around any really black cloud which has ragged bits underneath it is rarely time wasted. In Britain there are several tornadoes of varying intensity each year; they mostly occur in the Midlands. The central core may be about 200 yards across, in which the wind can reach perhaps 450 knots.

A dust devil is simply the visible sign of a strong thermal lift off. Dust devils in Britain are usually too small to do much more than blow some dirt around, but in hot dry countries they may be powerful enough to shift a parked aircraft, or produce unpleasantly sharp turbulence on take-off or during the approach. The spiralling motion is easy to see, but the direction of rotation can be turned into a subject for endless discussion; is it always clockwise, or anticlockwise?

Thunderstorm winds

The passage of a single thunderstorm can make a considerable local difference to the wind strength and direction. The storm moves along in the main wind stream, but usually a little faster and not always parallel. The extra speed comes because the cold air downfall at the rear of the storm, which is of higher pressure than that generally prevailing, is pushing under the rising air ahead (Fig. 9.2). The reason that the storm may not move directly parallel to the main wind stream is because its growth will be greater where it and the ground ahead are getting the full rays of the sun, and less where they are in shadow, and because new cells tend to develop on the flank, usually the starboard flank, of the storm. When a storm is in the vicinity its direction of movement should be determined to check whether it will pass harmlessly by, or whether it is moving, or growing, closer. If the storm is approaching the wind will slacken because the main wind is countered by air being sucked towards the storm to feed it. By the time that there is a lot of cloud overhead the wind may have dropped to calm. There is as yet no rain but the sky looks really threatening, and lightning is probably flashing away under the dark parts. A barometer may show a slight slow pressure fall up to this time, because the air locally is all going up. Suddenly – it always seems sudden – great gusts arrive, initially accompanied by blasts of dust or rubbish, and then rain or hail. The rain is made of big drops, and the hail may be very large. This is the cold air downfall. Pressure jumps and the wind really blows, with great risk to unsecured aircraft. From calm to gusts of perhaps 50–60 mph may take less than one minute.

As the storm passes on its way, the wind drops, and in due course the rain ceases. It does not stop as soon as the wind because, apart from other reasons for it continuing, it is falling from great heights and takes some time to reach the ground.

If a landing has to be made close to an approaching storm, severe turbulence as well as wind shifts, and altimeter error, must be expected right down to the ground. Taxiing a small aeroplane may be impossible without help; it is better to fly off to one side of the storm and wait in the clear air for the storm to pass by, and then land in the quieter air after it.

Sandstorms

In an ordinary wind sand will blow along over the surface; it is unpleasant but not dangerous, but when sand is picked up by convection and blows in a layer hundreds of feet deep, it is. In deserts dirt and sand blowing out from under a thunderstorm invariably precede the rain or hail, although the worst sandstorms are like line squalls, blowing at 40 mph or more through a depth of several thousand feet. A severe storm will quickly abrade a car or aircraft windscreen until it collapses, and completely strip the paint leaving only a thin shell of roughened metal. Parking a car tail to wind gives a greater chance of retaining a drivable vehicle – but not an aircraft unless the control surface locks are really tough.

Wake turbulence and jet blast

Even if we can avoid being blown over by natural storm and wind, there is no guarantee any longer that it will not be done to us by our own Big Brothers. In rather the same way as a boat produces a wake on the surface of the water, aircraft produce a pattern of turbulence behind them which may persist for several miles. It is primarily a pair of vortices – one behind each wing tip, which is due to the air under the wing tending to roll around the wing tip to the area of lower pressure on top. Flying across the flight path of an aircraft through these vortices will result in gusts up, down, down, up. Behind a small aircraft the turbulence can be felt but it is not hazardous. Behind a large one, however, the results may be lethal, as the small aeroplane can be rolled over with insufficient height to recover.

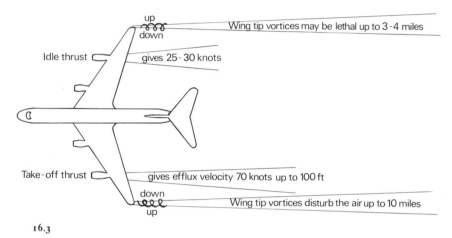

up
down
Wing tip vortices may be lethal up to 3-4 miles

Idle thrust — gives 25-30 knots

Take-off thrust — gives efflux velocity 70 knots up to 100 ft

down
up
Wing tip vortices disturb the air up to 10 miles

16.3

The power of the disturbance depends on the span, and the span loading; that is, the weight for every foot of span of the aircraft, and inversely on the speed; so a jumbo jet on the approach is the worst offender. Compared to the turbulence of the wingtip vortices, the effects of jets or airscrews, even at maximum power, is comparatively insignificant.

The vortices behind an aircraft decay only gradually over a period of several minutes; that is several miles, if the air is comparatively steady. Five miles should be regarded as the minimum distance when crossing the wake of a large slow-flying jet.

To reduce the chance of encountering severe wake turbulence the following should be taken into account.

1. The wake will move slightly downwards; we should therefore try to cross behind a large aeroplane at a higher level.

2. The lane of disturbance will move steadily downwind; when approaching to land behind a large aircraft on to a runway at an angle to the wind, the turbulence ahead can be avoided by keeping well on the upwind side of the straight glide path.
3. The large aircraft will normally make a low approach to the threshold; we are likely to have much more runway length available than we need, so can avoid the turbulence by making a much higher and steeper approach and landing well up the field.
4. Once the vortices come in contact with the ground they will spread outwards at about 5 knots or so, and can still be powerful after a couple of minutes, so if the wind is at all across the runway keep close to the windward side. If possible, wait.

Once airborne we should veer even further off the line of the runway.

Helicopters

A helicopter in forward flight creates two 'wingtip' vortices in a similar pattern to that of an aeroplane. When hovering, however, the downwash mushrooms out when it hits the ground, and this ring of turbulence will extend outwards for a considerable distance.

Jet blast

The main risk from jet blast is when an aircraft is parked within its influence, or when the large aircraft is taxiing out ahead. Jet blast is difficult to see on the ground except as a slight haze behind the aircraft, but a 747 produces a wind of 40 knots 1000 ft behind it. Any large jet on the ground with its beacon flashing should be considered to have its engines running until proved otherwise.

Landing if lost

If lost we can land in some funny places. The tests of initiative that
arise from ending up in farmers' fields, monastery gardens or State
highways are not the concern of this book, but we could have to
land in the sea or in the mountains on snow, and such emergency
fields may be considerably affected by the weather. Sometimes it
will require great skill to achieve a successful arrival, so we should
never add to the difficulties by also running out of fuel; having the
engine, even at reduced power, enormously reduces the risk of
failure.

Ditching

If we break cloud with only sea underneath, it may in due course
become necessary to ditch. If no land is in sight any ship is an
obvious source of succour, but just because we can see something
1000 ft long, we should certainly not assume that it has seen our
30-ft span. As soon as possible the aircraft should be flown over or
near the ship until it is obvious that it has been spotted. If visibility
is poor, and especially if it is felt that the lookout may not have
appreciated that there is an emergency, intermittent use of power
to indicate a failing engine will help. The landing should be made
at least half a mile ahead of the ship and a little to one side; if the
ship is large, long and fast, 2–3 miles ahead may not be excessive.
If the only ship is a yacht sailing, land *downwind* of it, about 400
yards in a fresh wind, and 100 yards or so in light airs.

The most important thing is to arrive on the water with the least
risk of either sinking or turning over, so the sea surface should be
studied carefully. There will be few problems if it is reasonably
smooth, but more often it will be confused with waves, wind gusts,
and the swell all going in different directions. Swell is the undulating
condition of the sea caused by a steady wind over a long period –
which maybe a very long distance away or some while ago. There
is a considerable length between the rounded crests which are

smooth and regular, and the greater the sea distance or *fetch*, from which the swell is coming, the longer will be the wave length. Waves are the undulations in the water surface travelling with the local wind, so there may be no correlation between their direction and that of the swell. The problem when ditching is to decide the best compromise line between the swell and the wind – just to land into wind without taking account of the swell can be disastrous. Equally,

Wind and sea	
Calm sea with no waves	0–10 knots
Scattered white caps	10–20 ,,
Many white caps	20–30 ,,
Streaks of foam	30–40 ,,
Spray from wave crests	40 knots plus

landing slap into the swell can be spectacular; the aircraft probably meeting the oncoming hump part way through the landing flare, and either being swamped or pitched up out of control. If landing downswell is considered, there may be a good flat length in which to land, since the swell is moving along in the same direction. Advantage should be taken of the situation to touch down just beyond a crest, so that the aircraft will stop before catching up with the next one. Unfortunately, the sea is often too steep to risk this manoeuvre.

The best direction to land is parallel to the swell, along either the trough or the crest – but on the back face and not on the arriving one. The United States Coastguard gives a guide as to what to do, as follows.

Wind 0–25 knots
Land parallel to the swell, choosing the direction that gives a headwind component.

25–35 knots
Select an intermediate heading, both into the wind and the swell. The higher the swell, the more the crosswind component that should be accepted, and the less the swell the more the aircraft should be headed into wind.

35 knots or over

In general, land into wind regardless of the swell, although in such strong winds the chance of both wind and swell going in the same direction is increased.

Wind direction and strength can be gauged by the wind streaks, which lie parallel to the wind. The 'white horses' are blown from the crests and fall downwind, with their foam being promptly run over by the waves. This gives the impression that the foam is sliding backwards and the illusion can be confusing (page 224, Beaufort Scale).

Having decided the direction in which the landing is to be made, the aircraft should be flown low over the water on a dummy run to check that the swell is not higher than anticipated; and then be brought in. The approach should be made with the aircraft being flown slowly low over the water, using plenty of power to hold the nose up, and this path should be followed until the water ahead looks smooth. The aircraft should then be lowered on to the water in a semi-stalled condition, care being taken to avoid either dropping it in from too high, or touching down too fast and being thrown into the air again. If the water is glassy it will be easy either to fly straight into it, or to land 50 ft or more up – again a steep power-on approach 7–10 knots above the stall, and continued right down on to the surface is best.

Down in the drink

We took off from Berck-sur-Mer at 1700 hrs GMT on 30 August 1970 bound for Southend. Shortly after reaching the mid-Channel point the engine suddenly lost virtually all power. As we were still more than 8 nautical miles from the English coast, and were losing height at about 1000 ft/min, I realised that, unless we could get the engine going again, we should have to ditch. At 2500 ft I (again) called Ashford, and found that we had descended sufficiently through the haze to be able to see a ship to our SW travelling away from us. I assessed that with the height available we should be able to overtake the ship, and ditch ahead of it.

We were both wearing life jackets, which was just as well because it would have been impossible to put them on in time and in the confined space of the cockpit. Mr Payne took our dinghy on to his lap so that it would leave the aircraft more or less automatically as he went through the door. We kept our headsets on to maintain communication with

—continued

Ashford control. We also released the door latches and Mr Payne put his foot in his door so as to keep it ajar. We opened the windows to allow the ingress of water and help equalise the pressure so that the doors would open more easily. We had no engine to 'blip' in order to attract the attention of the ship so I dived across the deck, converting my speed to height again having passed overhead, and then used this to glide as far ahead of the ship as possible, which turned out to be about $\frac{3}{4}$ mile.

The sea was very calm, but by looking along the sun path, which was about $45°$ on our starboard side, I was able to judge how high we were. I estimated that we were dead into a wind of 4–5 knots. I held off in a nose high attitude for as long as possible, progressively applying flap until, with the tail wheel in the water and full flap, the aircraft dropped in, and immediately stood on its nose.

The impact was severe and the cockpit went instantly under water. The windscreen shattered and the water rushed in with such force that it tore the dinghy pack out of Mr Payne's hands into the back of the aircraft. We did not see it again.

The aircraft started to sink immediately. I was aware of bubbles, then light green water turning to darker green. I could feel the pressure increasing in my ears. I do not remember undoing my lap strap but I do remember finding that my door had jammed shut. I assume that I broke the window with my elbow because this was later found to be lacerated, and I can recollect being stuck when partly out of the aircraft. I do not remember being particularly worried by not being able to breathe. Then I was free and swam to the surface. The tail was sticking straight up and the whole of the front part up to the trailing edge was under the surface. I could see my companion's jacket through the rear part of the canopy and thought he was trapped inside. Then I saw his head on the far side of the aircraft and remembered that he had not been wearing his coat. He was all right. The aircraft then sank having floated for no more than 20 seconds.

—From an account by F. A. Quick

Once the first contact has been made pilot skill ceases to be a significant factor, particularly if waves have damaged the tail surfaces; irregular deceleration should be expected, with big jolts. The shape of the aeroplane will have a bearing on the success or otherwise of the ditching. High wing machines will promptly sink at least up to the armpits, whereas the low wing aircraft will be more likely to sit on top, at least for a little longer. It goes without saying that the wheels should be retracted if this can be done.

We should obviously get out of the aircraft with the minimum delay, but stay with it. Rescue aircraft or ships will see the ditched machine or even a patch of oil more easily than a small human head, and even a few hundred yards to shore may be a much more difficult swim than is at first imagined. In any case, cold is the real enemy, and swimming dissipates body heat faster than staying still.

Lake ditching

In some countries, such as Finland, it may be preferable to land a glider on a lake instead of on forest or rough ground. A glider usually floats well and is not much harmed by a brief immersion in fresh water, so the risk is small. The technique is to land near where the lake is edged with reeds and a sandy shore, and be blown into them by the wind. In light winds it is possible to land downwind, jump out as the glider runs into the reeds and pull it ashore before it is more than superficially wet.

Landing on mountain snow

The most likely reason for having to land in the mountains is a combination of cloud cutting off the escape valley combined with the inability of the aircraft or the pilot to climb through cloud to the top. Getting lost and running out of daylight or fuel is another good reason for having to park on the heights. If a landing does have to be made the surface is most likely to be snow or glacier ice, and to be sloping. It is extremely unlikely that a satisfactory landing can be made on a downslope even when heading into a strong wind. If the wind is light it should be discounted and the landing made on the largest, smoothest, flat or upsloping area available. If the wind is strong we not only have to find an upslope with a suitable head-wind component, but one reasonably free of turbulence; we should fly above the intended landing place, and make an assessment of the direction and strength of the wind and work out how it is burbling around the mountain. Some idea can be obtained from blowing snow and from the way in which birds are flying – they like to avoid turbulence too. When we have made a plan we should fly experimentally into the region on a dummy approach, allowing margins of both speed and height – and with an escape route in mind should the aircraft unexpectedly hit a powerful down. While doing this a check should be made of the drift, and also of the state of the surface snow. Does it look new and soft, old and hard with protruding rocks, or windblown with ripples and alternate patches of hard and soft snow? If it seems to be smooth and well packed, or is a hard and empty ski piste, it might be worth lowering the undercarriage, but if there is any doubt at all the wheels should be left up. The approach to the slope should be made at a slightly higher speed than usual and this speed maintained until the last moment, in order that there will be enough in hand to flare out through what may be twice the normal angle. Unless the snow is very hard the aircraft will decelerate rapidly, but before it stops it must be turned or swung across the slope, so that it will not roll or slide backwards off the mountain. It is not particularly likely that the snow on which we land will carry an avalanche hazard – 22° of slope is needed – but there may be a risk that one will be brought down from above due to our rumbling around the mountains. The circumstances in which avalanches occur are when really heavy snowfalls have very recently put more snow on to the slope than it can reasonably hold, or when new thaw water runs under old snow; higher than usual temperatures should be considered a risk, and landing places under masses of overhanging snow should be absolutely avoided.

The pilot said . . .

. . . he had been flying the helicopter up a sloping valley to spot mountain sheep. Encountering gusty winds and turbulence he started to turn back, but as he crossed a ridge a downdraught forced him down on the mountainside. Winds were 15 knots with gusts to 30.

Having got back to Mother Earth, even if upside down in a snowdrift, the temptation to immediately trek off for help should be resisted unless the way down is obvious, easy, and unmistakable – despite visions of a cosy bar. Mountains can be very cold at night even in summer. If night is close help may be unlikely before morning, so snow should be banked up around the fuselage, particularly on the windward side, to provide insulation and the cabin made as habitable as possible for the night. At first light anything colourful should be spread out and a large SOS trodden in the snow. It's nice to be able to make it.

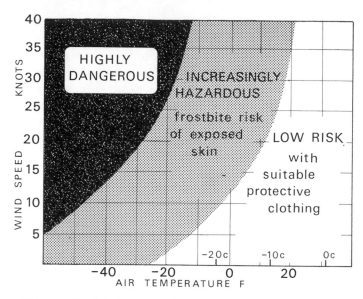

17.1 Wind chill risk for stranded pilot.

Find shelter; even lying flat on the ground helps.
Essential to protect head and neck. Keep dry.

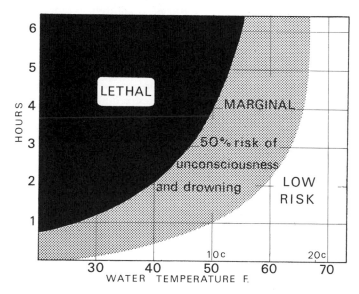

17.2 Wind chill risk for immersed pilot in ordinary clothes.

In the water discard only heavy coats or boots, keep on light close
weave clothing. Reduce movement to the minimum. If windy keep
body submerged and protect head and neck.

Appendices

"The clouds are green because they're trees."

Appendix 1

Static

While the generation of static electricity on a car is merely tedious, on a gas balloon it can be a disaster. Static will be developed when the air is dry, or the 'structure' has a low moisture content, and a charge is induced by some external work. This could be gas escaping from an orifice, or the folding or rubbing of fabric. All fabrics when free of moisture are likely to become electrically charged, but this is most noticeable with synthetic fibres because they do not hold moisture within the individual filaments. Although nylon and terylene are both considerably stronger than cotton they are not normally used for hydrogen balloons because of the greater risk of static; even so, cotton fabrics will be prone to 'spark' if torn, or if the rip is pulled, when the relative humidity is less than 50%. Some of the heat generated by the friction goes into evaporating moisture from the material, but if this is very dry sufficient surplus heat may be available to raise the temperature to the hydrogen ignition point. If this happens the balloon may catch fire, or it may explode, depending on the amount of air in the gas mixture. The initial fire may not be noticed for some seconds, since hydrogen burns with an invisible flame. The risk of disaster by static can be reduced by wetting the balloon and particularly the orifice, during deflation, and by the crew removing any nylon or other synthetic fibre garments.

Appendix 2

The Tephigram

In the book there have been several diagrams showing changes of temperature with height; to indicate, for example, the extent of an inversion. The Tephigram is a more sophisticated means of doing the same thing, but with the added advantage that it can be used to work out such things as cloud thickness, and not merely record old information. The name Tephigram is compiled from T indicating air temperature, and phi (Greek ɸ), which is used to signify *entropy* – approximating to the heat energy of the air.

The Tephigram plotting graph gives temperature in degrees C against entropy (and the dry adiabatic lapse rate). Superimposed on this basic grid is a pressure scale, which also gives height in feet and metres. The other curved lines running across the grid are for when the air is saturated (the saturated adiabatic lapse rate lines) and the amount of water vapour that can be contained in the air at any given temperature. Fig. A shows the relative positions of the lines for each of the above factors. It also shows a diagrammatic plot.

Fig. B shows an enlarged part of a tephigram with an 'actual' ascent drawn in. The information which is plotted on the chart is obtained from ascents by balloon or aircraft. On this chart the line ABCD represents the temperature recorded during the ascent plotted against pressure (and height). From A to B the temperature is seen to have fallen at the dry adiabatic lapse rate. At point B the temperature has reached the point at which condensation occurs; cloud begins to form at this level and due to the latent heat of condensation (page 18) the temperature falls at the saturated adiabatic lapse rate. However, at a somewhat greater height the aeroplane encounters an inversion. This means that the cloud reaches the level at which the surrounding air is now as warm, or warmer, than itself; no further cloud growth upwards is possible and the cloud either spreads out horizontally or dies.

At a greater height still the aeroplane thermometer shows that it has climbed through the inversion, point D, and that the temperature is once more falling at the dry adiabatic lapse rate.

As a simple practice exercise we can try to find the height at which cumulus will develop. First of all we need to know the pressure, temperature and humidity. Let us say that the pressure on the ground is 1000 mb, the temperature 15°C, and the relative humidity 70%. We plot the pressure and temperature (and call it point A). From the Water Vapour Content pecked lines it can be seen that point A corresponds to 11 gm

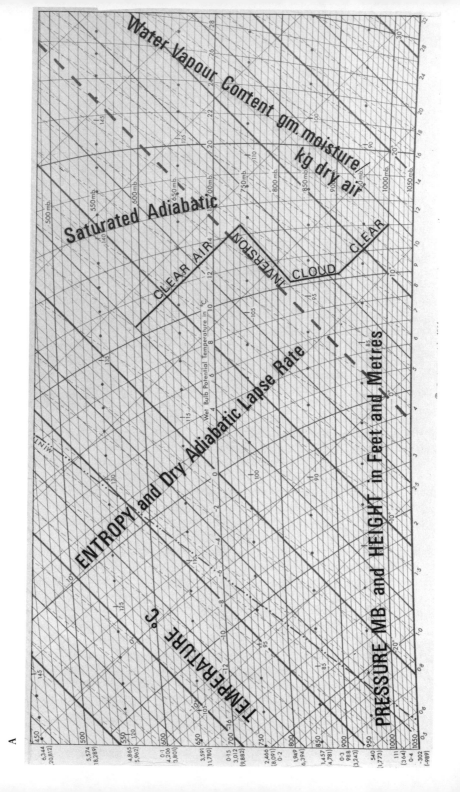

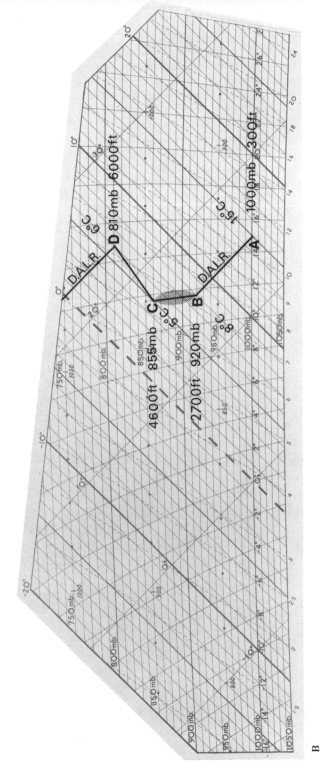

B

moisture per kg dry air. If the relative humidity on this day is 70%, then the air at the ground actually holds 7·7 gm moisture per kg dry air. To estimate cloud base draw a line from point A parallel to the dry adiabatic lines until the 7·7 gm pecked line is reached. This point will be found to occur at the 925 mb level. 925 mb is approximately equal to 2500 ft, and cloud base should be expected at this height. It should not, however, be expected that cloud base will stay at this height. It will change as and when the pressure temperature and humidity change, but as long as the basic data is available, the tephigram can be used to supply further information.

Appendix 3 Density Altitude

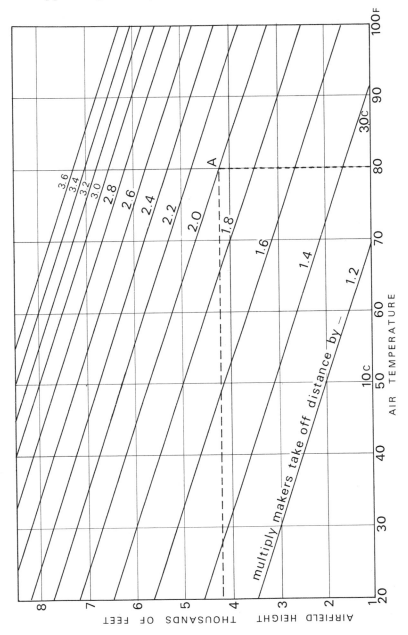

As airfield height and air temperature increases the take-off run will lengthen and the rate of climb worsen. To use chart: take manufacturers take-off distance and multiply by factor cutting height and temperature. E.g.: If the height above sea level of the airfield is 4200 ft and the air temperature 80°F the take-off run will double (A). It will, of course, be further increased by long grass, upslope, heavy load etc.

Appendix 4

Symbols, signs and abbreviations used on US Aviation Weather Reports

1. Sky and Ceiling

○ Clear: less than 0·1 sky cover.
◑ Scattered: 0·1 to less than 0·6 sky cover.
◍ Broken: 0·6 to 0·9 sky cover.
⊕ Overcast: More than 0·9 sky cover.
— Thin (when prefixed to the symbols).

— × Partial obscuration: 0·1 to less than 1·0 sky hidden by precipitation or obstruction to vision (bases at surface).
× Obscuration: 1·0 sky hidden by precipitation or obstruction to vision (bases at surface).

2. Letter preceding height of layer identifies ceiling layer and indicates how ceiling height was obtained. Thus:

A Aircraft.
B Balloon.
D Estimated height of cirriform clouds on basis of persistency.
M Measured.
R Radiosonde or radar.

W Indefinite.
U Height of cirriform layer, ceiling layer unknown.
/ Height of cirriform layer, nonceiling layer unknown.
V Varying ceiling.

3. Visibility. Reported in statute miles and fractions: V variable.

4. Weather and Obstruction to vision symbols.

A Hail	F Fog	RW Rain showers
AP Small hail	GF Ground fog	S Snow
BD Blowing dust	H Haze	SG Snow grains
BN Blowing sand	IC Ice crystals	SP Snow pellets
BS Blowing snow	IF Ice fog	SW Snow showers
D Dust	K Smoke	T Thunderstorms
E Sleet	L Drizzle	ZL Freezing drizzle
EW Sleet showers	R Rain	ZR Freezing rain

5. Precipitation intensities are indicated:

— — Very light; – Light; (no sign) Moderate; ⓒ Heavy.

The intensity indicator *follows* the precipitation symbol.

6. Station level pressure, corrected to sea level, indicated by final 3 digits in mb.

7. Temperature and dewpoint. Reported in Fahrenheit. When the spread is reduced to 3°F or less fog is possible.

8. Wind. Direction in tens of degrees from true North, speed in knots, 0000 indicates calm. G=gusty. Peak speed of gusts follows G, or Q when squall is reported. WSHFT followed by local time group indicates occurrence of wind shift.

E.g. 1825G35 WSHFT1615=Wind 180° 25 knots gusting 35 knots, wind shift at 1615 hours.

9. Weather information abbreviations:

ABNML	Abnormal	IPV	Improve
ABV	Above	IR	Icy runway
ADVN	Advance	JTSTR	Jet stream
ADVY	Advisory	KOCTY	Smoke over city
ADQDS	All quadrants	LTGCCCG	Lightning, cloud to cloud,
AMD	Amend		cloud to ground
APCH	Approach	LTLCH	Little change
ARPT	Airport	MDFY	Modify
BINOVC	Breaks in overcast	NCWX	No change in weather
BLZD	Blizzard	OBSC	Obscure
BRKN	Broken	OCFNT	Occluded front
CAT	Clear air turbulence	OTLK	Outlook
CAUFN	Caution advised until further	OVC	Overcast
	notice	PCPN	Precipitation
CAVU	Clear weather, vis. < 10 miles	PIREP	Pilot report of weather
CB	Cumulo-nimbus	PRESFR	Pressure falling rapidly
CIG	Ceiling	PSR	Packed snow on runway
CVR	Cover	RAFL	Rainfall
DLA	Delay	RAREP	Radar weather report
DRFT	Drift	RH	Relative humidity
DRZL	Drizzle	RVR	Runway visual range
DTRT	Deteriorate	SLR	Slush on runway
DWPNT	Dewpoint	SNFLK	Snowflake
EXPC	Expect	SQLN	Squall line
EXTRM	Extreme	SVR	Severe
EXTSV	Extensive	TDA	Today
FCST	Forecast	THDR	Thunder
FLG	Falling	THSD	Thousand
FQT	Frequent	TOVC	Top of overcast
FRMN	Formation	TRML	Terminal
FROPA	Frontal passage	TRRN	Terrain
FRZ	Freeze	TURBC	Turbulence
GRAD	Gradient	TWRG	Towering
HAZ	Hazard	U or	Unlimited, unrestricted,
HDWND	Headwind	UNKN	unknown
HLSTO	Hailstones	UWNDS	Upper winds
HRZN	Horizon	VLNT	Violent
HURCN	Hurricane	VSBY	Visibility
ICGIC	Icing in clouds	WB	Weather Bureau
INSTBY	Instability	WR	Wet runway
INTMT	Intermittent	WX	Weather
		YDA	Yesterday

Appendix 5   International weather vocabulary

English	Danish	Dutch	French	German	Italian	Spanish
Altitude	Altitude	Hoogte	Altitude	Höhe	Altitudine	Altura
Anticyclone	Anticyklon	Hogedrukgebied	Anticyclone	Antizyklon	Anticiclone	Anticiclon
Airfield		Vliegueld	Aéroport	Flughafen	Aeroporto	El aeropuerto
Amendment	Forbedring Aendring	Verandering	Changement Amendement	Anderung	Correzione Emendamento	Enmiende Rectificacion
Area	Areal	Gebied	Zone, région	Gebiet	Area	Zona
Backing	Drejer Til Venstre Drier mod Solen	Krimpend	Recul du vent	Krimpen	Rotazione Sinistra	Rolanda a la Izquierda
Beaufort	Beauforts	Windschaal van Beaufort	Echelle de Beaufort	Beaufortskala Beaufort Windstarke	Scala di Beaufort	Escala de Beaufort
Calm	Vinstille	Windstil	Calme	Windstille	Calma	Calma
Centre	Centrum Center	Centrum	Centre	Zentrum	Centro	Centro
Clouds	Skyer	Wolken	Nuages	Wolken	Nubi, Nuvole	Nubes
Clouds (broken)	Gebrokkent	Gebroken	Nuages fragmentés Troué	Aufgerissen Durchbrochen	Rotto, nubi	Quebrado, nubes Fragmentadas
Cloudy	Over skyet	Bewolkt	Nuageux	Bewölkt	Nuvoloso	Nublado, Nuboso
Coast	Kyst	Kust	Côte	Küste	Costa	Costa
Cold	Kold	Koud	Froid	Kalt	Freddo	Frio

English	Danish	Dutch	French	German	Italian	Spanish
Cyclonic	Cyklonisk	Cycloonachtig Cycloonisch	Cyclonique	Zyklonisch	Ciclonico	Ciclonica
Deep	Dyb	Diep	Profond	Tief	Profondo	Profundo
Deepening	Uddybende	Verdiepend	Creusement	Vertiefung	Approfondimento	Ahondamiento
Dense	Taet, Tyk	Dicht	Dense	Dicht	Densa	Denso
Depression	Lavtryk	Depressie	Dépression	Tiefdruckgebiet	Depressione	Depresion
Direction	Direktion Retning	Directie	Direction	Richtung	Direzione	Direccion
Dispersing	Adsprendende	Verstrooiend	Déperdition	Zerstreuung	Dispersione	Disipacion
Drizzle	Stovregn	Moxtregen	Bruine	Sprühregen	Spruzzatore Pioviggine	Llovizna
Dusk	Tusmorke	Schemering	Brune crépuscule	Dämmerung	Crepuscolo	Crepusculo
East	Øst	Oosten	Est	Ost	Est	Este
Extensive	Udstrakt	Uitgebreid	Etendue	Verbeitet	Estesa	General
Falling	Faldende	Vallend	En baisse	Fallend	In diminuzione	En disminucion
Filling	Udfyldende	Vullig	Comblement	Auffüllen	Riempimento	Relleno
Fine (fair)	Skon, Klar	Mooi	Clair, Beau	Schönwetter Heiter	Sereno	Sereno
Fog	Taage	Mist	Brouillard	Nebel	Nebbia	Niebla

English	Danish	Dutch	French	German	Italian	Spanish
Forecast	Vejrforundiegelse	Weers-verwachting	Prévision	Vorhersage	Previsione	Prevísión
Formation	Formation	Formatie	Formation	Bildung	Formazione	Formacion
Forming	Formende	Vorming	Développement	Formation Bildung	Sviluppo	Formante
Frequent	Hyppig	Veelvuldig	Fréquent	Haufig	Frequente	Frecuenta
Fresh	Frisk	Fris	Frais	Frisch	Fresco	Fresco
Front	Front	Front	Front	Front	Fronte	Frente
Front (passage of)	Forbifart	Frontpassage	Passage d'un front	Front Durchzug	Passaggio di un fronte	Paso de un frente
Frost	Frost	Vorst	Gelée	Frost	Brina	Escarcha
Gale	Hård Blaest Stormende Kuling	Storm	Coup de vent	Stürmischer Wind	Burrasca	Viento Duro
Gale warning	Stormvarsel	Stormwaarschuwing	Avis de coup de vent	Sturmwarnung	Avviso di burrasca	Aviso de mal viento
Good	God	Goed	Bon	Gut	Bene, buono	Bueno
Gust	Vindstod Vindkast	Windstoot Windulaag	Rafale	Windstoss, Böe	Colpo di vento Taffica	Rafaga, Racha
Gusty	Stormfuld Byget	Buiig	(Vent) à Rafales	Göig	Tempestoso	En Rafagas en Rachas
Hail	Hagl	Hagel	Grêle	Hagel	Grandine	Granizo

English	Danish	Dutch	French	German	Italian	Spanish
Haze	Dis	Nevel	Brume sèche	Dunst (Trockener)	Foschia	Abundante, Violento
Heavy	Svaer	Zwaar	Abondant	Schwer	Pesante, Violento	Huracan
Hurricane	Orkan	Orkaan	Ouragan	Orkan	Uragano	Aumentar
Increasing	Stigning	Stijgen	Augmentant	Zunehmend	In aumento crescente Aumentare	
Intermittent	Intermitternde	Intermitterend	Intermittent	Zeitweilig	Intermittente	Intermitente
Isobar	Isobar	Isobar	Isobare	Isobar	Isobara	Isobara
Isolated	Isolere	Isoleren	Isole	Einzeln	Isolato	Aislado
Latitude	Bredde	Breedte	Latitude	Breite	Latitudine	Latitud
Light, slight	Tynd, let	Licht, Gering Zwak	Faible	Schwach	Leggero, Debole	Ligero, Dehil
Lightning	Lyn	Bliksem	Eclair	Blitz	Lampo	Relampago
Line Squall	Bygelinie	Buienlÿ	Ligne de grain	Böenfront	Linea della Tempesta	Linea de Turbonada
Local	Lokal	Plaatselijk	Local	Örtlich, Lokal	Locale	Local
Longitude	Laengde	Lengte	Longitude	Länge	Longitudine	Longitud
Low	Lavtryk	Laag, depressie	Bas	Niedrig, tief	Basso, depressione	Bajo, depresion
Meridian	Meridian	Meridiaan	Meridien	Meridian Langenkreis	Meridiano	Meridiano

English	Danish	Dutch	French	German	Italian	Spanish
Mist	Let Täge	Nevel	Brume légère	Dunst (Feuchter)	Caligirie	Neblina
Moderate	Middelmadig Moderat	Matig, Gematigd	Modéré, Réduit	Mässig, Mittel	Moderate Discreta	Moderado, Regular
Moderating	Beherske Aftagende	Afnemenol	Moyenment Décroissant	Nachlassend	Medianente Che diminuisce	Medianente Disminuyendo
Morning (in the)	Om Formiddag Om Morgenen	Voormiddag	Le matin	In Morgen	Al or il Mattina	Por la Manana
Moving	Bevaegende	Bewegend	En déplacement	Bewegend	In Movimento	Movimiento
North	Nord	Noorden	Nord	Nord	Settentrionale	Septentrional Boreal
Occasional	Tilfaeldig	Toevallig	Occasionnel	Gelegentlich	Occasionale	Occasional
Occlusion	Okklusion	Okklusion	Occlusion	Okklusion	Occlusione	Occlusion
Off-shore wind	Fra land vind	Van land wind	Vent de Terre	Ablandiger Wind	Vento di suolo	Viento Terral
On-shore wind	I land vind	Aan land wind	Vent de mer Brise de mer	Anlandiger Wind Seewind	Vento di mare	Viento de Mar
Overcast	Overtrukket	Betrokken	Couvert	Bedeckt	Coperto	Cubierto
Period	Periode	Tijdvak, Periode	Periode durée	Periode	Periodo	Periodo
Period of validity	Gyldighed periode	Periode van geldigheid	Periode de validité	Gultigkeitsdauer	Periodo di validità	Periodo de validez
Poor	Ond, dårlig	Slecht	Mauvais	Schlecht	Scarso, Cattiva	Mal, Mala

English	Danish	Dutch	French	German	Italian	Spanish
Precipitation	Nedbor	Neerslag	Précipitation	Niederschlag	Precipitazione	Precipitación
Pressure	Tryk	Druk	Préssion	Druck	Pressione	Presión
Quickly	Kvick, Rask	Snel	Vite	Schnell	Rapidamente Velocemente	Pronto
Rain (continuous)	Regn Uagbrudt	Regen Onafgebroken	Pluie Continue	Regen Anhaltend	Pioggia Continua	Luvia Continuo
(Slight) rain	Tynd, Let	Licht, Gering Zwar	Faible	Leicht	Leggera, Debole	Debil Legero
Ridge	Ryg	Rug	Dorsale, Crête	Rücken, Kammlinie	Dorsale, Cresta	Dorsal, Cresta
Rising	Stigning	Rijzend	En hausse Monter	Steigend	Ascendente, In salita	Ascendente, Subir
Rough	Oprort	Ruw	Agitée	Stürmisch	Agitato	Bravo o alborotado
Scattered	Sprede, Strø	Verspreiden	Dispersée	Zerstreut	Diffuso	Difuso
Sea Breeze	Sø Bris, Havbris	Zeesbries	Brise de mer	Seebrise	Brezza di mare	Virazon, brisa de mar
Shower	Byge	Bui	Averse	Regenschauer	Rovescio	Aguacero, Chubasco
Sleet	Slud, Sne og Regne	Natte sneeuw	Neige et pluie mêlées, Neige fondue	Regenschnee	Nevischio Pioggia Ghiacciata	Aguanieve
Slowly	Langsom	Langzaam	Lentement	Langsam	Lentamente	Lentemente
Smooth	Glatte	Vlak, glad	Embellie	Glatt	Tranquillo Calmo	Tranquilo Calmo

English	Danish	Dutch	French	German	Italian	Spanish
Snow	Sne	Sneeuw	Neige	Schnee	Neve	Nieve
South	Syd	Zuiden	Sud	Süd	Sud	Sur
Squall	Byge	Windvlaag	Grain	Böe	Tempesta Groppo	Turbonada
Stationary	Stillestaende Stationaer	Stationair	Stationnaire	Stationär Stillstehend	Stazionario	Estacionario
Storm	Uvejr storm	Storm	Tempête	Sturm	Tempesta Tembrale	Temporal
Strong	Staerk Kraftig	Sterk Krachtig	Fort	Stark Kräftig	Forte	Fuerte
Thunder	Torden	Donder	Tonnerre	Donner	Tuono	Trueno
Thunderstorm	Tordenvejr	Onweer	Orage	Gewittersturm	Temporale	Tronada Tempestad, Borrasca
Thermal		Thermick	Thermique	Thermisch	Convezione termica	Termal
Time	Tid	Tijd	Temps	Zeit	Tempo	Hora
Trough	Udlober	Trog	Creux	Tiefdruckrinne	Saccatura	Vaguada
Variable	Foranderlig Variabel	Veranderlijk	Variable	Veränderlich	Variabile	Variable
Veering	Drejer til Hojre	Ruimend	Virement Virage	Rechtsdrehend	Destrogiro	Dextrogiro
Visibility	Synsvidde	Zicht	Visibilité	Sicht	Visibilità	Visibilidad

English	Danish	Dutch	French	German	Italian	Spanish
Warm	Varm	Warm	Chaud	Warm	Caldo	Calido
Waterspout	Skypumpe	Waterhoos	Trombe marine	Wasserhose	Tromba marina	Tromba marina
Wave (form)	Bolge	Golf	Vagues	Welle	Onda, ondulatorio	Onda
Weather report	Vejr-Melde	Weerbericht	Météorologique Bulletin	Wetterbericht Wettermeldung	Rapporto	Informe, Aviso Boletin
West	Vest	Westen	Ouest	West	Ponente	Oeste
Whirlwind	Hvirvelvind	Wervelwind	Tourbillon de vent	Wirbelwind	Turbine	Trobellino
Wind	Vind	Wind	Vent	Wind	Vento	Viento

Appendix 6

Conversion Tables

Kms		Statute Miles
1·61	1	0·62
3·22	2	1·24
4·83	3	1·86
6·44	4	2·48
8·04	5	3·11
9·66	6	3·73
11·26	7	4·35
12·87	8	4·97
14·85	9	5·59
16·09	10	6·21
32·19	20	12·43
40·23	25	15·53
80·47	50	31·07
160·92	100	62·14

Nautical miles		Statute miles
0·87	1	1·15
1·74	2	2·30
2·60	3	3·45
3·47	4	4·61
4·34	5	5·76
5·21	6	6·91
6·08	7	8·06
6·95	8	9·21
7·82	9	10·36
8·68	10	11·51
17·37	20	23·03
21·71	25	28·79
43·42	50	57·58
86·84	100	115·15

Metres		Feet
0·30	1	3·28
0·61	2	6·56
0·91	3	9·84
1·22	4	13·12
1·52	5	16·40
1·83	6	19·68
2·13	7	22·97
2·44	8	26·25
2·74	9	29·53
3·05	10	32·81
6·91	20	65·62
7·62	25	82·02
15·24	50	164·04
30·48	100	328·08

Square metres		Square feet
0·09	1	10·76
0·19	2	21·53
0·28	3	32·29
0·37	4	43·06
0·46	5	53·82
0·56	6	64·58
0·65	7	75·35
0·74	8	86·11
0·84	9	96·88
0·93	10	107·64
1·86	20	215·28
2·32	25	269·10
4·64	50	538·20
9·28	100	1076·40

Litres		Inp. Glns
4·55	1	0·22
9·09	2	0·44
13·64	3	0·66
18·18	4	0·88
22·73	5	1·10
27·28	6	1·32
31·82	7	1·54
36·37	8	1·76
40·91	9	1·98
45·46	10	2·20
227·30	50	11·00
454·60	100	22·00

US Glns		Imp. Glns
1·20	1	0·84
2·40	2	1·66
3·60	3	2·50
4·80	4	3·33
6·00	5	4·16
7·21	6	4·99
8·41	7	5·83
9·61	8	6·66
10·81	9	7·49
12·02	10	8·33
24·02	20	16·65
60·05	50	41·63
120·09	100	83·27

Kg		Pounds
0·45	1	2·20
0·91	2	4·41
1·36	3	6·61
1·81	4	8·82
2·27	5	11·02
2·72	6	13·23
3·17	7	15·43
3·63	8	17·68
4·08	9	19·84
4·54	10	22·05
9·07	20	44·09
11·34	25	55·12
22·68	50	110·23
45·36	100	220·46

Metres		Yards
0·91	1	1·09
1·83	2	2·19
2·74	3	3·28
3·66	4	4·37
4·57	5	5·47
5·49	6	6·56
6·40	7	7·66
7·32	8	8·75
8·23	9	9·84
9·14	10	10·93
18·28	20	21·86
22·85	25	27·33
45·70	50	54·66
91·40	100	109·32

Millibars	Inches of mercury
1026·1	30·30
1022·7	30·20
1019·3	30·10
1015·9	30·00
1013·2	29·92*
1012·5	29·90
1009·1	29·80
1005·7	29·70
1002·3	29·60
* Standard atmosphere.	

See also page 36.

Index